Gustavo Jimenez

Fatigue testing machine redesign and automation

AF304165

Gustavo Jimenez

Fatigue testing machine redesign and automation

Rotary bending fatigue machine

ScienciaScripts

Imprint

Any brand names and product names mentioned in this book are subject to trademark, brand or patent protection and are trademarks or registered trademarks of their respective holders. The use of brand names, product names, common names, trade names, product descriptions etc. even without a particular marking in this work is in no way to be construed to mean that such names may be regarded as unrestricted in respect of trademark and brand protection legislation and could thus be used by anyone.

Cover image: www.ingimage.com

This book is a translation from the original published under ISBN 978-620-0-41478-6.

Publisher:
Sciencia Scripts
is a trademark of
International Book Market Service Ltd., member of OmniScriptum Publishing Group
17 Meldrum Street, Beau Bassin 71504, Mauritius
Printed at: see last page
ISBN: 978-620-3-60752-9

Copyright © Gustavo Jimenez
Copyright © 2021 International Book Market Service Ltd., member of OmniScriptum Publishing Group

Redesign and automation of Rotating Bending Fatigue Machine

CONTENTS

Introduction

One of the fundamental requirements of the design procedure for mechanical parts that will form part of a device is knowledge of the material from which they will be made. A material is identified by a set of macroscopic and microscopic properties. The process by which these properties are measured and specified is called characterisation and, depending on the macroscopic or microscopic properties to be studied, there are specific procedures for their determination.

From a macroscopic point of view, the following properties of materials are known: mechanical, magnetic, electrical, optical properties.

Mechanical properties include: modulus of elasticity, yield strength, tensile strength, percentage elongation and reduction in area, toughness, hardness, fatigue strength.

On the other hand, in order to carry out a material characterisation process, it is necessary to have the appropriate technological infrastructure to obtain accurate and reliable information. These technologies must, in turn, be on a par with technical advances in the manufacture and commissioning of equipment, specifically equipment for testing materials. However, the economic circumstances of some countries in the world, especially those known as third world countries, which are characterised by a lack of capacity to purchase modern equipment, make it necessary for them to adopt alternatives to make up for the deficiencies in their technological infrastructure. One of these possible strategies is to modernise existing machinery, whenever this is feasible.

In some cases, it is cost-effective to make a transformation to existing equipment; this is achieved through detailed studies of its functional characteristics, the design of the improvement proposal, procurement of the set of parts and pieces that make up the design, and finally, its testing and commissioning.

The Universidad Politécnica Territorial del Estado Aragua "Federico Brito Figueroa" (UPTA) located in the city of La Victoria, has in its facilities a rotary bending fatigue testing machine that is inoperative. The reasons for this situation are: damage to the revolutions counter and also that the application of the load is carried out manually, which can lead to inaccuracy in the process. As the machine is not working, the main effect is that the programmes of the curricular units Materials Technology and Design of Mechanical Elements are not being fulfilled.

Scientific Problem

Based on the above, the following scientific problem is posed as a scientific problem: In the current conditions of the UPTA rotary bending fatigue testing machine, particularly with the engine speed counter system damaged together with a load application system that may be inaccurate, it is not possible to carry out fatigue tests for teaching or research purposes for multiple disciplines.

Hypothesis

The creation of a load application system on the UPTA rotary bending fatigue testing machine that operates accurately, without the risk of overstressing specimens, together with an automated data collection system, would ensure that it operates efficiently, would make it similar to modern fatigue testing equipment and would contribute to the improvement of the delivery of the Mechanical Design and Materials Technology programmes.

Object

UPTA rotary bending fatigue testing machine

Field

The field of application covers mechanical design, mechanical behaviour of materials and their characterisation.

General objective

Redesign the load application system and data acquisition method of the rotary bending fatigue machine to perform the test automatically.

Specific objectives

1. Review the current literature on fatigue phenomena, control systems and data acquisition.
2. Design the mechanism for applying the load to the specimen to be tested.
3. Design the measurement and data acquisition system.
4. Test, by means of simulation and cold runs, the system for applying bending loads to the specimen.
5. Test data acquisition programmes.

Research tasks.

Fulfilment of the objectives set out above depends on the implementation of the following activities:

1. Collect theoretical information relevant to the phenomenon of rotary bending fatigue, as well as information on the conduct of the relevant

test.
2. Collect technical information on existing devices to perform the test, which are similar or related to the one under study.
3. Generate the model, with the aid of CAD tools, of the bending load application system.
4. Simulate the designed models using CAE tools.
5. Select the electromechanical components to be used in the application of the bending load on the specimens.
6. Designing the interfaces of the data acquisition system
7. Test test data generation routines and test control and monitoring routines.

Scientific Novelty

The scientific novelty presented in this work is related to the way in which the bending load is applied. This action is carried out from the control system that is part of the test supervision and data capture interface, avoiding having to apply the load manually as in the original system. Additionally, the software application gives the user the possibility to choose the type of model with which to plot the cantilever beam rotary bending fatigue test data.

The contributions or impacts generated with the transformation of the current machine are:

Socially:

It allows the application of this project in other universities, contributing modestly to the motivation of the student population to pursue and successfully complete their studies in Mechanical Engineering.

In academia:

It allows fatigue practices to be carried out, thus strengthening the student's training in the areas of knowledge related to materials and mechanical design.

In science:

It allows researchers to carry out part of the fatigue characterisation of some type of material under study, provided that it is susceptible to this type of test. It allows the verification of any hypothesis put forward by them.

Financially:

It contributes indirectly to the strengthening of the local productive apparatus.

Scientific Methods Employed

The deductive-analytical method is used in this work. Through the deductive process, the mathematical models by which the control program operates are generated and the test data are processed. With the analytical method, the entire test system, i.e. the machine, is broken down into subsystems, which allows a better understanding of its operation and facilitates the redesign work.

Justification.

Professional training processes, such as a Mechanical Engineering course or programme, will be more efficient to the extent that up-to-date technical resources are available to contribute to their delivery. Traditional subjects such as Materials Technology and Mechanical Design include laboratory practices in their programme contents that reinforce the topics taught in the hours devoted to theory.

In the specific case of fatigue theory of materials, it is appropriate for students and also researchers to have up-to-date equipment for testing metal specimens under fatigue conditions.

With the implementation of the machine transformed into a mechatronic device that allows the acquisition of fatigue test data under rotary bending conditions and, at the same time, allows the storage of the data obtained in this test, the teacher will have an updated tool and at the same level as the latest generation machines; this will allow him to prepare a more intuitive course of materials or design, which allows the student to know in depth the fatigue behaviour of the material he wishes to select for a particular design or to characterise it.

This work can also serve as an initial idea for similar transformations in other laboratory equipment. In fact, in the specific case of UPTA, there is already an existing tensile testing equipment that has undergone modernisation, which reinforces the fact that if there is no capacity to purchase sophisticated equipment, it will be possible, to the extent that this is feasible, to remedy the situation by converting existing equipment.

Chapter 1. Theoretical Framework.

1.1 Introduction

The fatigue test is part of the alloy characterisation process. With this, the fatigue life of the part made with the material under study can be determined. The fatigue phenomenon has been the subject of research on several occasions as can be seen in [2], [3], [4], [5], [6] as it is very complex in nature and its effects can be devastating if it is not considered at the time of the design of mechanical components. A mechanical part will show fatigue behaviour depending on the state of loads to which it is subjected, depending on its geometry (particularly if it has stress concentrators), as well as the discontinuities in the composition of the material and the environmental conditions in which it is working.

The part, while in operation within the mechanical assembly to which it is attached, cannot normally be examined to determine whether fatigue failure may occur; this result, unfortunately, is only visible when the failure event and subsequent damage to the equipment - which may include means of transporting humans, animals, various materials including explosives, etc. - has already occurred. In the past, this was thought to be a supernatural event or due to structural changes as expressed in [7].

Over the course of time, the nature of fatigue in the materials of which the parts that make up the various mechanical assemblies are made was progressively understood. Thus, parameters were established for carrying out experiments to determine the fatigue life of the material of the mechanical parts to be designed; test equipment has evolved and today we have advanced technological devices [7], [8], [9].

The phenomenon of fatigue failure occurs in well-defined stages. Lee [6], states that the fatigue failure process involves the following stages:

a) Crack core formation (nucleation)
b) Short crack growth
c) Long crack growth
d) Final fracture [6].

Similarly Bangoura [2] mentions the stages of crack nucleation, crack growth, crack propagation and finally fracture. It is clear then, that for fatigue failure to occur, cyclic loading and stress concentrators must be present, in the vicinity of which cracks form.

It is considered that fatigue analyses can be grouped or classified into three (3) types:

a) Stress-based analysis and design
b) Deformation-based analysis and design
c) Frequency domain analysis [6].

1.2 The phenomenon of fatigue. General

As previously expressed, for fatigue failure to occur, the part must be under the action of fluctuating loads, whose value may be below the value of the yield strength of the material [2] and by the presence of discontinuities in the composition of the material or by changes in the geometry of the part [10], [11], [12].

It is estimated that 80 % of machine failures are due to fatigue [12]. The term **"fatigue"** attributed to this phenomenon is due to an analogy with human fatigue; it was first used in 1839 by Poncelet. Even at that time, the phenomenon was not fully understood and when observing the fracture plane in ductile materials it was believed that it had crystallised, "tired" and become brittle due to the oscillation of the loads, and although Wohler (1819-1914) [13] demonstrated that the parts of broken pieces (in that case, the axles of railway carriages) were as strong and ductile in tests of

stress as the original part, the term fatigue is still used to identify failures due to the action of time-varying loads.

1.2.1. Dynamic loads. Characteristics and types

In the different elements that make up a machine, two basic types of loads or stresses that act on them can be distinguished, depending on whether or not they vary over time. Loads that remain unalterable over time or whose variation is very slow are called **static loads,** and those that do show variation over time are called **dynamic loads**. Static loads are also called **stationary loads**, and dynamic loads are called **cyclic loads, non-stationary loads** or **transient loads** [7].

The graphical representation of these types of loads can be seen in figure 1.1:

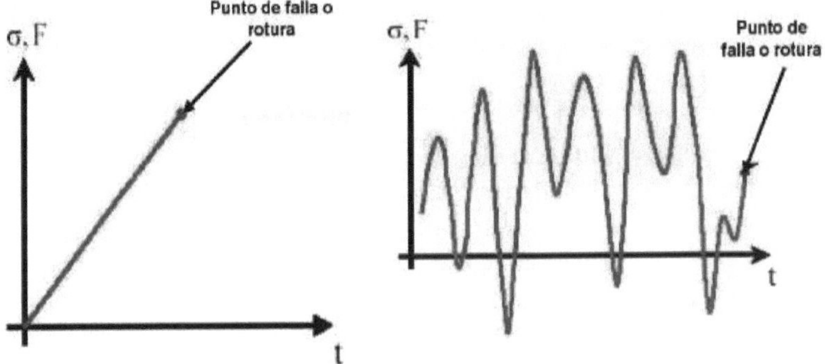

Figura 1.1. Tipos de cargas o fuerzas que actúan en un elemento de máquina: a) Cargas estáticas o estacionarias y b) Cargas dinámicas o no estacionarias. Fuente: [1]

Several configurations of dynamic loads can be derived from figure 1.1.b, however, the one used to describe the fatigue phenomenon will be the one that is simplest in its mathematical expression. The mathematical model(s) describing this phenomenon will be discussed later in this chapter.

1.2.2. **Stages of the fatigue failure process**

Fatigue cracks start at several sites simultaneously and propagate when one flaw dominates and grows faster than the others [2]. As discussed earlier in this chapter, fatigue failure occurs in three stages: crack nucleation, crack growth and crack propagation leading to fracture of the component.

1.2.2.1. Nucleation stage. Engineered materials are generally inhomogeneous and anisotropic. This is due to the material's own inclusions left over from the manufacturing process. On the other hand, the part, for design reasons, has a particular geometry which may be composed of notches, diameter reductions, notches among other elements; both these geometric characteristics and microscopic imperfections operate as stress concentrators.

Due to the variation of the forces acting on the part, and as a consequence of this, the variation of the stresses generated, local plastic creep can occur due to stress concentration, even when the nominal stress in the section is well below the elastic limit of the material [2]. Slip bands are created or appear at the crystallised edges of the section, and as the stresses alternate, more bands and thus more microscopic cracks appear. Pre-existing voids or inclusions will serve as stress risers for crack initiation. A crack will tend to form more quickly in a brittle material than in a ductile one, since in the former the creep phenomenon does not occur, unlike in the latter.

1.2.2.2. Crack propagation stage.

After the crack has formed, the mechanisms of fracture mechanics become apparent. There is a sharp crack which, with the alternation of stresses, forms a plastic zone that reduces the stress concentration and produces a slight growth; if the stresses vary from a minimum to a maximum, but at a value well below the elastic limit, the crack closes, momentarily ceasing the plastic creep, then the crack becomes sharp again, acquiring a larger dimension [2]. It should be noted that the crack is formed by the tensile component of the alternating stress.

1.2.2.3 Fracture stage.

Bangoura [2] states: "The crack will continue to grow as long as cyclic tensile stresses and/or surface corrosion factors are present. At some point, the crack size becomes large enough to raise the stress concentration factor K at the crack tip to the level of the fracture toughness of the material Kc, and in the next tensile stress cycle sudden failure occurs instantly. This failure mechanism is the same, regardless of whether the situation K=Kc has been reached by crack propagation because the nominal stress has not been raised sufficiently" [2]. The part continues to deteriorate due to crack growth and the net cross-section of the part is so reduced that it is incapable of resisting the load from a static point of view and fatigue failure occurs [4].

Figure 1.2 shows an example of a specimen that failed due to fatigue. For more information on fatigue fracture, see reference [14].

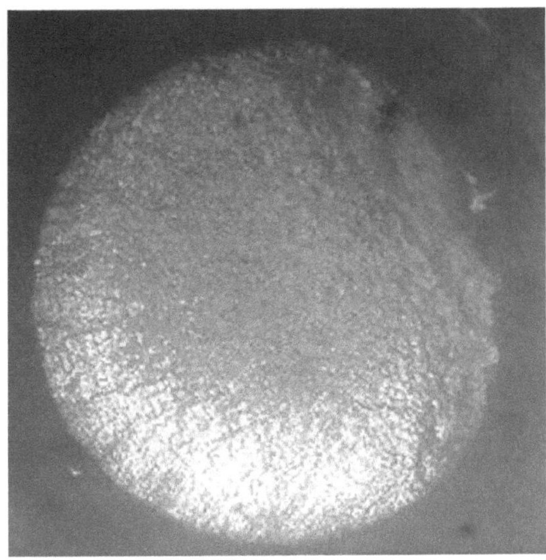

Fig. 1.2. Fracture plane of specimen subjected to rotary
bending fatigue test.
Source [1]

1.2.3. **Types of cyclic stresses.**

The cyclic stresses generated in a part can be classified or grouped into axial (tensile or compressive), bending or torsional. It is desirable to express these stresses in the simplest possible terms.

A general expression could be as follows [7]:

$$\sigma(t) = A * sen(C * t) + B$$

(1.1)

Where,

- $\sigma\,(t)$: alternating stress as a function of time
- *A*: amplitude
- *B and C*: constants

11

Or it could be more complex [7]. In any case, the following notable stresses can be found [7]:

- Maximum voltage: denoted by $S_{máx}$

- Minimum voltage: denoted by $S_{mín}$

Its expression depends on the type of stress.

The two are related to obtain other stresses that are of importance in the study of fatigue. These are:

The average stress, which is denoted by:

$$S_m = \frac{S_{máx} + S_{mín}}{2} \tag{1.2}$$

The voltage amplitude, defined as:

$$S_a = \frac{S_{máx} - S_{mín}}{2} \tag{1.3}$$

Range of stresses:

$$S_r = S_{máx} - S_{mín} \tag{1.4}$$

Stress ratio: the ratio of the minimum to maximum stress.

$$R_s = \frac{S_{mín}}{S_{máx}} \tag{1.5}$$

Amplitude ratio: the ratio of the voltage amplitude to the mean voltage.

$$A_a = \frac{S_a}{S_m} = \frac{S_{máx} - S_{mín}}{S_{máx} + S_{mín}} = \frac{1 - R_s}{1 + R_s} \tag{1.6}$$

Depending on the values in equations (1.2) to (1.6), the stresses can be presented in four different ways:

- Symmetrical, $S_m = 0$, $R_s = -1$

- Intermittent, $S_{mín} = 0$, $R_s = 0$

- Constant, $S_a = 0$, $R_s = 1$

- Pulsing, 0< Rs>1

- Alternate, -1< Rs>1.

All these stress classes are valid for both tensile and compressive, normal and shear stresses.

The graphical representation of these types of fluctuating voltage is shown in Figure 1.3.

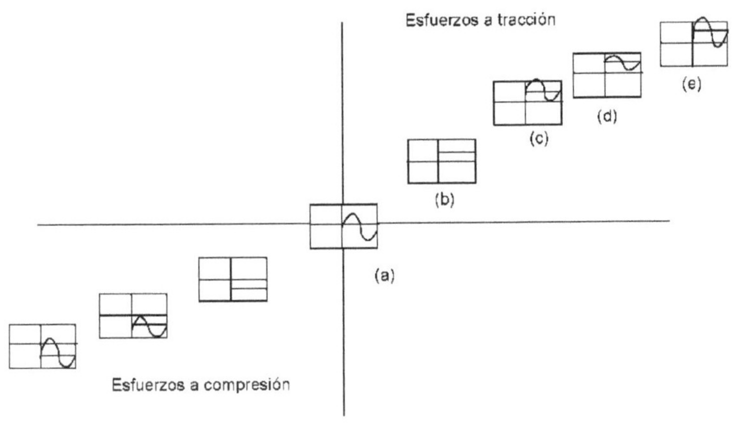

Fig. 1.3. Types of cyclic stress: (a) Symmetric, (b) Constant, (c) Intermittent, (d) Pulsating, (e) Alternating. The effect in tension and compression is shownSource
: Author

1.2.4. **Types of fatigue tests. Mathematical models for data processing.**

In the study of fatigue, several types of tests have been implemented, generally associated with the types of cyclic stresses mentioned in the previous section. The rotating beam test is the most widely used [15], [8], [9], [14], [16], [17], [18]; [19]; for this reason, this type of test will be dealt with in the present work, in addition to the fact that the machine to be redesigned is of this type.

Fatigue tests with rotary bending are performed with two types of machines: the Moore specimen testing machine [16], [20], [15], [4], [1], [18], [19] and the cantilever rotary bending machine [17], [8], [9], [21]. Moore's machine, in addition to the machine shown in [21], uses a set of weights to apply the bending load, while the models presented in [16], [8], [9], [18] use a dynamometric system to apply the bending load on the specimen. Models of both machines are shown in figure 1.4 and 1.5.

Fig. 1.5.
Wrapped beam fatigue machine.
Source [9]

Fig. 1.4. Moore's fatigue machine.
Source [20]

If both images are compared, it can be seen that the cantilever test machine is more compact than the Moore test machine; this gives it an advantage, and that is that it does not require much space to be located in the laboratory. The dimensions of the machine shown in figure 1.5 can be found in [9]. This machine is very similar to the one that will be redesigned and automated in this work.

1.2.4.1, **Mathematical models for data processing.**

The data generated in the rotating beam tests can be processed with various mathematical models. These are usually regression equations where the relationship between the stress at which the specimen fails (*S*) and the number of cycles to failure *(N)* is expressed.

These equations are used to obtain the fatigue life curve of the specimen. These curves are called *S-N* curves. The most frequently used regression models or equations for metals correspond to the following criteria [22], [14], [6].

1.2.4.1.1. **Effort - Life Model:**

This model corresponds to the first of those used to study the phenomenon of fatigue in metals. At the end of the 19th and beginning of the 20th century, the complexity of industrial machines increased, and with this complexity came an increase in the number of failures due to the action of cyclic loads. During the test, following this model, the specimen is subjected to cyclic loading until failure. The stresses generated in this test are those expressed in equations (1.1) to (1.6) [6]. The typical equation used to represent the fatigue life, according to this approach, is given by

$$S_a = S'_f * (2N_f)^b$$

$$(1.7)$$

Where,

S_a: amplitude of efforts

S'_f: Fatigue resistance coefficient

b: fatigue strength exponent. Its values range from -0.05 to -0.12 for most metals.

Equation (1.7) is known as the Basquin Equation. The applicability of this analysis criterion extends to the cases of bending stress, axial stress and torsional shear stress. Further equations for the Life-Stress criterion can be found in reference [6]. A graph obtained with this model is shown in figure 1.6.

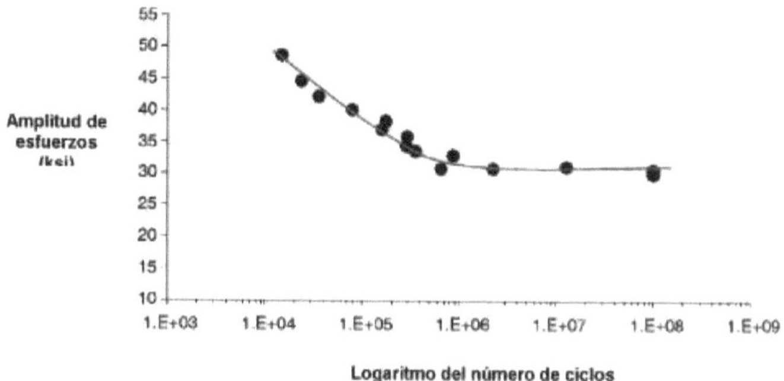

Fig. 1.6. Fatigue curve according to the Life-Stress
criterion. Source [6]

1.2.4.1.2. Life - Strain Model: In this model [3], in a situation of low number of cycles - in which the deformations occur within the elastic range - the relationship between the life of the component and the amplitude of stresses is defined by the Basquin equation

$$\frac{\Delta S}{2} = S'_e * (2N_i)^b$$

(1.8)

In which,

- $2N_i$. number of load reversals

- S'_e. Fatigue strength coefficient

- b. fatigue strength exponent, the most common values of which are given in equation (1.7); c. fatigue resistance exponent, the most common values of which are given in equation (1.7)

For high load levels, where elastic deformation predominates, the number of cycles to failure of the component is linked to the strain amplitude by the expression

$$\frac{\Delta \varepsilon}{2} = \varepsilon'_f * (2N_i)^c$$

(1.9)

Where,

- $s'f$: fatigue ductility coefficient

- c: fatigue ductility exponent, varies from -0.5 to -0.7

Both models act as asymptotes in the logarithmic-double representation of the S - N curve. By means of the correction through the Mason-Coffin-Morrow model, it is proposed to fit the prediction curve to both asymptotes, thus achieving a valid model for any load state.

$$\frac{\Delta \varepsilon}{2} = \frac{S'_e}{E} * (2N_i)^b + \varepsilon'_f * (2N_i)^c$$

(1.10)

The graph of this model is shown in Figure 1. 7

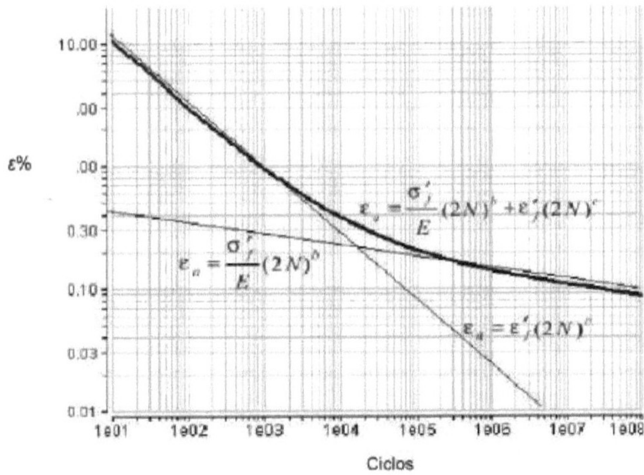

Fig. 1.7. Life versus deformation. Source [3]

I.2.4.2. **Other models for fatigue testing**: It is of interest to the author to include in the present work, other models for fatigue life determination, such

17

as the Ripoll's Spectral Stress Field and Cumulative Damage model [5] and the prediction model of Castillo et. al [23].

In [23], the authors consider that the Life-Stress model is recommended or, at least, complementary due to the complexity of other analysis calculations that are inherent to fracture mechanics. The authors propose a general mathematical model, which includes a probabilistic definition capable of representing the different characteristics of the Life-Stress curves (S-N curves) of different materials that, for example, may or may not present a fatigue or endurance limit (stress range below which fatigue failure does not occur).

This model is chosen because it meets the conditions outlined for its application in [23].

The proposed equation is as follows:

$$E(\log N; \log \Delta S) = 1 - e^{-\left[\left(\frac{(\log N - B)*(\log \Delta S - C)}{D} + E\right)^A\right]}$$

(1.11)

In which,

N: fatigue life measured in cycles

AS: range or amplitude of efforts

E *(logN; logAS)*: probability of failure distribution function

C: fatigue or endurance limit, represented by a horizontal asymptote (Figure 1.6).

B: threshold value or limit of the number of cycles, represented by a vertical asymptote
A: shape parameter of the Weibull distribution
D: scaling parameter of the Weibull representation function

E: parameter that fixes the position of the limit or zero probability curve.

The last three parameters A, D and E are related to the parameters of the normalised Weibull distribution.

The graph of the Castillo et al. model is shown in Figure 1.8.

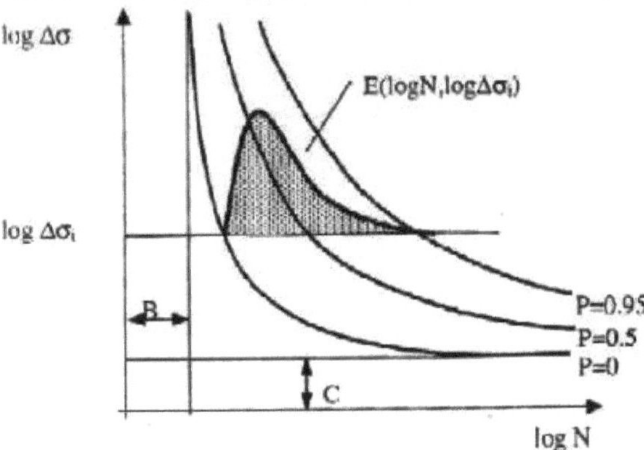

Fig. 1.8. Model of Castillo et al. Source [23].

In the model proposed by Ripoll, the author proposes the following [5]: "...a probabilistic model that serves as a basis for the fatigue analysis of metallic materials subjected to any load spectrum in any load range (tension, compression or mixed) to provide a theoretical and practical response to a field of research...".

widely explored before, but has not been able to obtain a mathematical expression involving statistical, physical and material behaviour concepts for the type of loading described above, so as to serve as a basis for the beginning of a new way of approaching the fatigue of materials based on a new model". To this end, the author reviews known fatigue models such as Basquin, Palmgren-Miner, Stromeyer, Haibach and others.

The proposed model is

$$p = 1 - e^{\left\{ -e^{\left[C_0 + C_1 S^*_m + C_2 S^*_M + C_3 S^*_m S^*_M + \left(C_4 + C_5 S^*_m + C_6 S^*_M + C_7 S^*_m S^*_M \right) \log N^* \right]} \right\}}$$

(1.12)

Where,

$C0$, $C1$, $C2$, $C3$, $C4$, $C5$, $C6$, $C7$ are the parameters defining the model; they are obtained by optimisation procedures.

$$S^*_m = S_m / S_0$$

$$S^*_M = S_M / S_0$$

$$N^* = N / N_0$$

S_m: S mínimo

S_M: S máximo

S0 and N0 are values of stress and number of cycles that are appropriately chosen to obtain the dimensionless parameters. In figure 1.9 (a) and (b), the graphs corresponding to the validation of the model with the alloys Steel 42CrMo4 and AlMgSi1 can be observed.

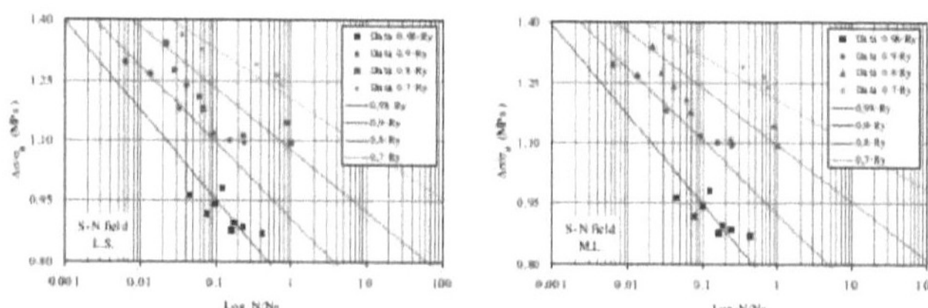

Figure 1.9 a: Resulting Wohler field for 42CrMo4 steel with the parameters obtained by the maximum likelihood method (right figure) and least squares regression (left figure). Source [5]

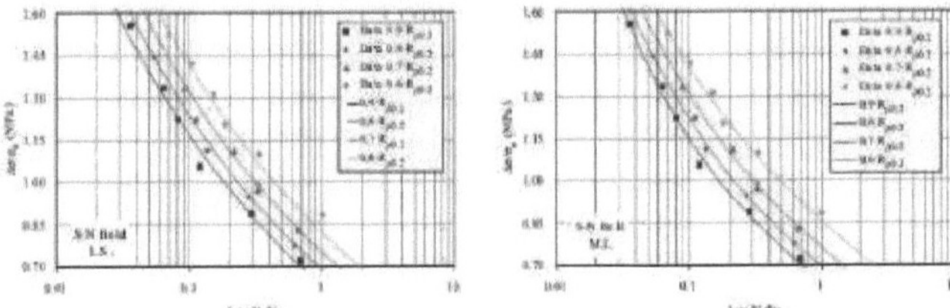

Figure 1.9 b: Resulting Wohler field for the AlMgSil alloy with the parameters obtained by the maximum likelihood method (right figure) and least squares regression (left figure). Source [5]

1.3. Automation of fatigue testing machines.

The continuous modernisation of society brings with it the implementation of ever more advanced technologies. Industrial processes, in accordance with this modernisation, have driven the development of technological tools to meet their proposed objectives in an efficient manner. Many, if not most, of these technologies are the result of the fusion of many disciplines that some time ago would have been considered impossible and unthinkable; there are also mass consumer products that are made under this same conception: digital cameras, computer hard drives, smart home systems, among others. All of them have in common that they are the product of the fusion of disciplines such as mechanical engineering, electronic engineering and computer science in a discipline known as *Mechatronics*. The term *Mechatronics* is used to describe the integration of microprocessor-based control systems, electrical systems and mechanical systems [24], representing the synergistic union of mechanics, electronics and computing [25], [26]. Other authors express that mechatronics is not a science but should be seen as a philosophy, a fundamental way of doing and seeing things that by their very nature require a unified approach in order to be created [27]. Under a mechatronics-based approach, it is possible to arrive at technological solutions that satisfy efficiency, performance, reliability and cost requirements [28]. In the author's opinion, mechatronics should not be seen as a panacea, but as a useful tool to execute precision tasks, as an extension of human capabilities. A mechatronic system is an alternative for improving industrial, medical and domestic processes, among others.

In this field, the materials characterisation process, specifically fatigue testing, also makes use of mechatronics to meet the objectives set. The cantilever beam type rotating bending fatigue machine models developed by the German company GUNT [9], the Spanish Edibon [8] and the model developed by Brandolisio *et al* [29] are available; the latter is a variant of the Moore machine driven by a servo-hydraulic system. In the former, the mechanical device is combined with an interface developed in LabView for the input of the test data. The model provided is shown in figure 1.5. The interface developed in the LabView software is shown in figure 1.10.

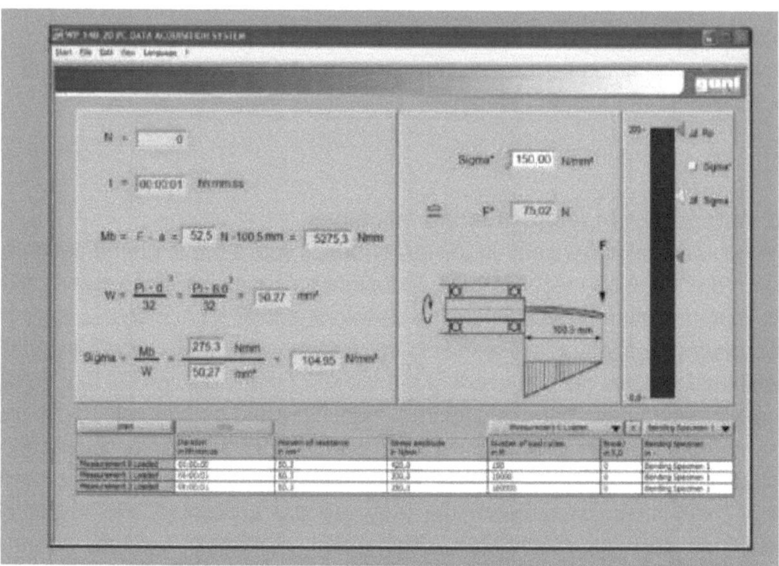

Fig. 1.10. Lab View interface for processing machine data produced by GUNT.
Source [9]

The system offered by Edibon uses a SCADA system for control, data acquisition and processing tasks. Its basic configuration is shown in figure 1.11.

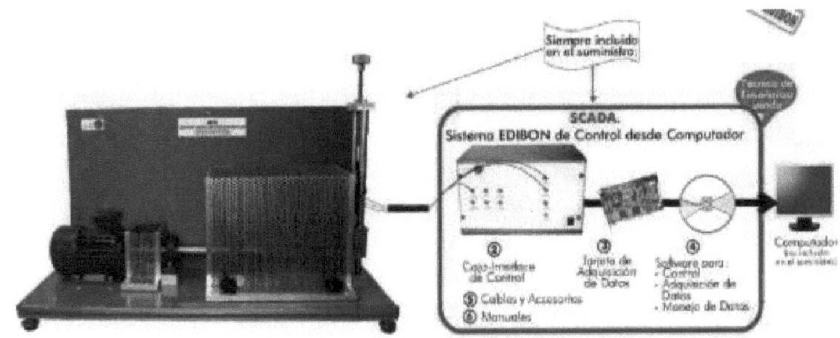

Fig. 1.11. Fatigue machine developed by Edibon. Source [8]

The prototype developed by Brandolisio *et al. is shown* in figure 1.12. This prototype is quite different from the one that will be used in this research.

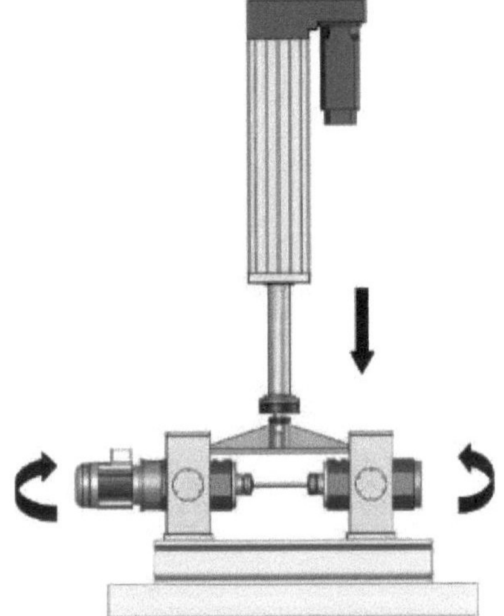

Fig. 1.12. Brandolisio machine. Source [28]

The models previously referred to in [8] and [9] have a common characteristic: the way in which the bending load is applied to the specimen is done manually, so they follow the same procedure for testing fatigue specimens in the known way. This procedure is described for this type of machine in [30]. The question arises: why not automate the way in which the bending load is applied in the cantilever type rotary bending machine; with the advances in mechatronics, this objective can be achieved and all the functions of the machine, including data processing, can be integrated in a

23

compact module which, in addition to fulfilling the function of generating the cyclic inversion of the stresses in the specimen, allows the data to be manipulated and provides information in accordance with the mathematical models set out in previous sections.

1.4. Servo motors and DC motors.

Among the components used in a mechatronic system are servo motors or servos. A servo motor is a device similar to a DC motor that has the ability to move to any position within its operating range and remain stable in that position [31], the servo has a shaft whose performance is controlled [32]. Servos are frequently used in radio control systems and robotics, but their use is not limited to that, due to their low inertia, torque transmission capability, approximately 3 kg.cm for a standard servo which makes them quite strong for their size. It also has proportional power for mechanical loads, therefore, a servo has reduced power consumption [31]. The basic composition of a servo is a motor, a gearbox and a control circuit [31], [33] as shown in Figure 1.13. Most servos have a 180° output shaft travel but can be modified to operate as a motor.

Fig. 1.13. Parts of a standard servomotor: 1-Motor, 2-Control board, 3-Reducing gears, 4-Output end element. The large black parts are the housing components. Source [32]

The position control is carried out internally by the servo via a potentiometer which is mechanically connected to the output shaft and controls an internal pulse width modulator (pwm) and compares it with the external pwm input of the servo via a differential system and thus changes the position of the output shaft until the values are equal, so that the servo stops at the specified position. In this position, the servo motor stops consuming current and only a small current flows to the internal circuit. If the servo is forced (by moving the output shaft by hand), at this point the internal differential control detects this and sends the necessary current to the motor to correct the position [34]. A servo is controlled by the application of a pulse of specified duration and frequency. Servos have three connection wires: one for Vdc voltage, one for Gnd ground, and a third that receives the control pulse train, which will cause the internal differential control circuit to move the servo to the position indicated by the pulse width [34]. Depending on the manufacturer, servos have the following characteristics

pulse duration ratings and cable layout as shown in table 1.1

Table 1.1. Pulse types and wiring for different servos. Source [33]

Manufacturer	Pulse duration (ms)				IIIIIJ cable layout		
	min.	neutral.	max... I	Hz	+ batt	-batt	pwm.
Futaba	0.9	1.5	2.1	50	red	black	white
Hitech	0.9	1.5	2.1	50	red	black	yellow
Graupner/Jr	**0.8**	1.5	2.2	50	red	brown	orange
Multiplex	1.05	1.6	2.15		red	black	yellow
Robbe	0.65	1.3	1.95	50	red	black	white
Simprop	1.2	1.7	2.2	50	red	blue	black

It can be seen then, that a servomotor is of wide utility as an actuating element of a mechatronic system, its characteristics such as torque transmission capacity, dimensions and power consumption levels make it the ideal element for the fulfilment of the main objective of the present work. However, direct current motors or DC motors should be considered; these can also generate adequate torques for driving the proposed load application device, as can be seen in [35].

1.5. **Data acquisition and processing systems.**

For an electromechanical system to acquire the typical configuration of a mechatronic system, it is necessary for its structure to include a data acquisition and processing subsystem, which connects the physical electromechanical model and the specialised software. The latter is in charge of capturing the device's input signals (acquisition), analysing, processing and finally exercising the required control actions.

Signals (input and output) are physical variables or quantities measured in various parts of a system, which, when processed, generate the desired information [36]. In practice, there are many signals that can be found in various real systems, with the electrical signal, in the form of a current or voltage signal, being the easiest to measure because the processing of other non-electrical signals can be carried out using sensors and transducers that transform the original signal into an electrical signal.

Signals can be classified into two main groups: time-continuous (TC) and time-discrete (TD) signals, each of which can be either deterministic or random [36], [26]; the most common signals can be seen in Figure 1.14.

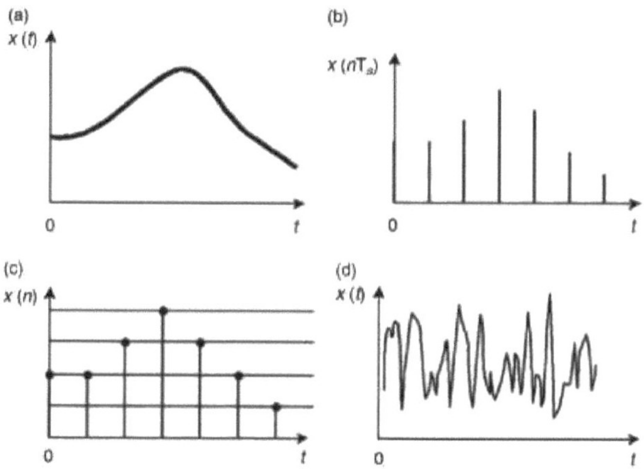

Fig. 1.14. Types of signals: a) Continuous, b) Sampling, c) Digital and d) Random. Source [34]

Deterministic signals can be expressed analytically, while the behaviour of random signals cannot be accurately predicted. A continuous signal in time has a specific value for each time value; the discrete signal has a unique value at discrete points, i.e., integer values of n in the time domain.

Deterministic signals can be placed in two different groups in relation to the repeatability of their values or not in time; we have periodic signals and non-periodic signals. The first group satisfies the relation $x\ (t)= x\ (t+T)$, $-\infty < t < \infty$. An example of this type of signal is the sine function, whose expression is $x\ (t)= A^*sen\ (\Omega^*t + \theta)$ [36], where θ is the phase angle in radians, $\Omega=2^*\pi^*F$ is the frequency in rad./s and F is the frequency in Herz.

In the case of a discrete periodic function, the signal is defined, in generic terms, by the expression $x\ (n)= x\ (n + N)$ with $-{}^{TM} < n < M$, where N represents the period. The mathematical foundations of the functions used as signals are described in detail in [36], [37] and [38].

The purpose of a data acquisition system is to capture and analyse some kind of real-world phenomenon. Light, temperature, pressure and torque are some of the many different types of signals that can be linked to a data acquisition system. This, in turn, can simultaneously produce electrical signals which can either intelligently control mechanical systems or produce a stimulus, which the data acquisition system can measure its response [26].

In the data acquisition and processing stages, the signals that are commonly manipulated are temperature, displacement and speed, among others, which are analogue or continuous (see figure 1.14). So, in order for the computer, using the appropriate software, to be able to process them, they must be previously treated. This task of transformation from analogue to digital type is called analogue - digital conversion of the signal, and the devices in charge of this change are called analogue - digital converters so that later, when the system has processed the information, it returns the response in analogue format and it is then said that a digital - analogue conversion has taken place. This demonstrates, to this author's knowledge, that for effective computer control a conversion from a continuous or analogue signal to a discrete or digital signal is necessary.

1.6. Partial conclusions

The following conclusions can be drawn from this chapter:
1. Fatigue failure of a mechanical part is a phenomenon that, regardless

of the advances made in materials technology, is always present and must be a property that must be known in depth, which is why it is necessary to test samples of the material under study; such tests must be carried out with suitable technological resources. Judging by the bibliographic material consulted, the study and mathematical modelling of the phenomenon of fatigue failure in rotary bending has not changed and all the works consulted that deal with this phenomenon are based on research carried out in previous years.

2. The mathematical models used to study fatigue in metallic materials are of the exponential and linear-logarithmic type. With the help of software tools it is possible to implement applications that process, through these models, the data of the number of cycles and stress amplitude from the fatigue tests by rotating bending of the cantilever beam type.

3. The servomotors and DC motors referred to in this chapter are suitable for use as actuators because, despite their small size, they develop torques and forces of considerable magnitude. Their position control and motor speed control tasks can be carried out with the same application created to process the data generated in the test.

Chapter 2. Proposed solution

2.1. Introduction

The implementation of appropriate methods and machinery for fatigue testing involves the generation and commissioning of various systems such as those shown in the previous chapter. The tendency, as we have seen, is to use technical resources that minimise human intervention as much as possible; this implies the use of technologies in which automatic operation is the fundamental function. Chapter 1 showed that appropriate technical systems can be achieved through the use of mechatronics. This discipline encompasses both theoretical and technical aspects to meet the demands of fatigue testing.

In this chapter, a brief description of the current machine (located at the UPTA facilities) is given, as well as a description of the proposed system; comparisons between the two are shown. The load application system is described and its choice is justified.

The mechanical design packages that can be used for the design of the alternative solution are mentioned and briefly described.

The applications, to generate the test control and monitoring system, are also presented. These are LabView and Movicon. An overview of the interface to the test monitoring and control system is also given.

2.2. Current system

The current system is a semi-automatic machine. The load is applied manually, as well as the motor drive and the resetting of the rev counter, while the automatic part is in the switch-off of the motor by means of a detector.

The current model is shown in figure 2.1, which outlines the main parts of the model [17].

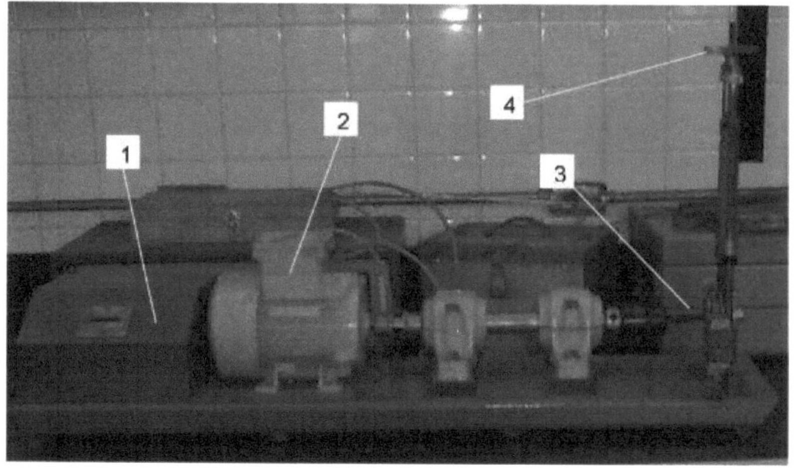

Figure 2.1: UPTA Fatigue Testing Machine
Main components: 1- Tachometer, 2- Motor, 3- Specimen holder, 4- Load application mechanism.
Source: [17]

It is of interest to this author to replace components 1 and 4. Component number 1 - tachometer - is faulty and component number 4 - manually operated dynamometer - is often inaccurate. In [30], the procedure to perform the test with this type of device is explained and it is clearly warned not to apply too much load to the dynamometer, thus avoiding unexpected damage to the specimen and alteration of the data expected in the test.
Figure 2.2 shows another view of the load application system.

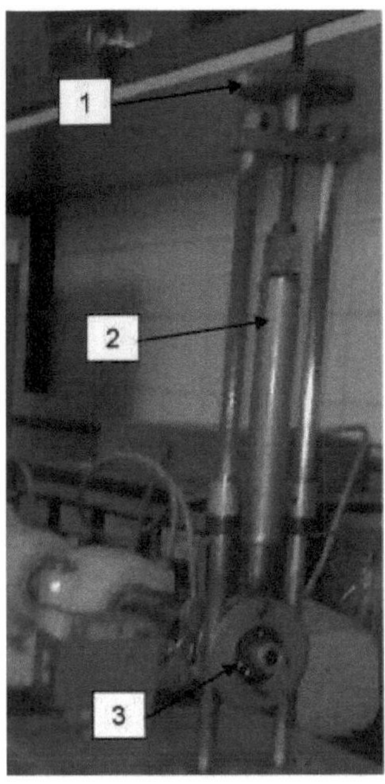

Figure 2.2: Load application system of the current equipment: 1- Actuating knob
, 2- Dynamometer, 3- Specimen holder
Source: The author

The switching off of the motor at the moment of specimen breakage is achieved by actuating the stop detector, which has a roller type follower. Figure 2.3 shows the detail of this device.

Figure 2.3: Detail of the engine stop detector of the current equipment.
Source: Author

In this case, the follower 1 detects the breakage of the specimen. This action is performed because the specimen itself offered, before breaking, a resistance to the displacement of the dynamometer column. After the specimen breaks, this resistance disappears, the column moves axially, together with the specimen support (marked with the number 3 in figure 2.2), and finally the follower actuates the valve. When this actuation takes place, the signal is sent to the motor via 2, which causes it to switch off and stop.
The WP140 model from the German company GUNT has this type of specimen breakage detector, while the EEFC model from the Spanish company EDIBON detects specimen breakage by means of a load cell connected to its built-in control module [8], [9]. In both models, the load is applied manually.

The WP140 model includes, as an option, a graphical interface for test data processing; it is therefore found that the machine manufactured by EDIBON is slightly more advanced than the German model.
The EDIBON machine is shown in Figure 2.4.

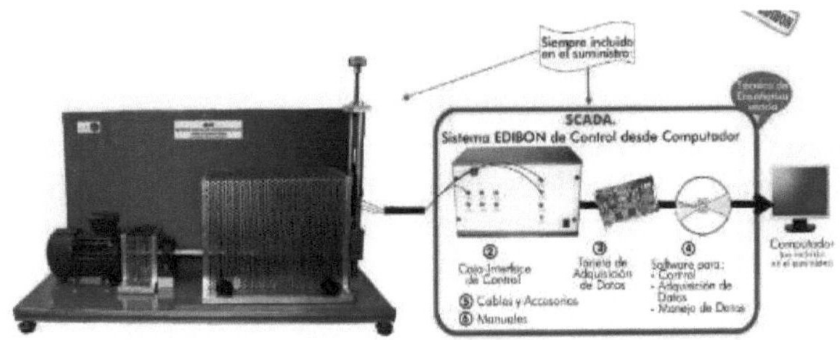

Figure 2.4. Fatigue testing machine produced by EDIBON. Source [8]

2.3. **Proposed system**

In the redesign of the machine proposed in this work, the application of the load is carried out in a software-controlled way, as well as the stopping of the engine. The test data storage function is added in order to obtain the corresponding curve - S-N or Wohler curve - to be included in the test report. The generation of such a report is not part of the functions of the proposed solution; the similarity of the proposed design with the Spanish machine (manufactured by EDIBON) is in the software control and in the detection of the applied load by strain gauge. The common feature that the proposed design alternative - potential feature - shares with the model developed in Germany is the use of LabView as a software tool for the control and monitoring of the fatigue test.

To detect the application of the load on the specimen as well as its breakage, the proposed device has a *strain gage* located on the support at the front end of the specimen, whose function is to detect and measure the value of the applied bending force. When the strain gage breaks, the sensor (strain gauge) registers the corresponding voltage variation and thus the control system orders the motor to be switched off. The assembly and subsequent removal of the tested specimen is done manually.

Based on the characteristics of the models presented in the previous section (test data management by software, use of LabView package, load application detection by strain gauges), the redesign proposal for the load application system is presented below. In contrast to [17], in which bending is performed with servo driven wedges, the proposed redesign of this machine takes advantage of the screw mechanism to generate the bending load.

While it is true that, as discussed in the previous chapter, servo motors have a high load application capability, the long test duration time - up to 550 hours at 3000 rpm of the current system - makes the current consumption considerable; additional mechanical devices are then required to ensure that the applied load is maintained, which leads to unnecessary complication of the mechanical part. Due to this fact and the fact that a servo motor never turns a full revolution, DC motors are preferable. The reference given in chapter 1 shows the performance of a commercial DC motor unit. Therefore, the solution is basically to drive and control the movement of the screw, while retaining the original structure of the current equipment.
The most suitable means of transmitting motion to the screw is by means of a gear transmission due to its particular characteristics. The specification of this is made considering the following aspects:

- The power required to generate the bending of the specimen.

- Geometrical complexity of transmission

- Feasibility

To estimate the power, the maximum working load is considered, which in this case is 300 N, given that this value was originally used in the UPTA fatigue tests, and the rotational speed. The latter is considered not to be very high, preferably as low as possible.

In terms of the rotational speed and the maximum torque (produced by the 300 N) the value, in kW, of the power is given by [39]:
Where,
T: Torque, in N-m
N: Engine rotational speed, in revolutions per minute (rpm) The torque modulus is given by the expression

$$P = \frac{T \times N}{9550} \qquad (2.1)$$

$$T = F_t \times r$$

(2.2)

In which,

F_t: Tangential force
r: Turning radius.

For the case of the present system, the diameter of the knob is 68 mm, (0,068 m) so the radius is 0,034 m, the tangential force F_t is 300 N. Applying (2.2), the required torque value is

T= 300x0, 034= 10,2 N-m.

To establish the speed, reference values are sought from DC motor catalogues.

Based on the condition that the rotational speed is as low as possible - and thus a low power requirement is also guaranteed - the DC motor referenced in Chapter 1 is chosen:

Model S330144.
Nominal voltage: 12V
Nominal speed at 12 V: 6 rpm
Standstill torque: 74.8 kg-cm (7.3 N.m)
Force: 15 kg/cm (1.5 N / m).

Therefore, the suggested speed is 6 rpm.
The complexity of the drive geometry is related to the dimensions of its components. This depends on the type of transmission to be used. It was previously stated that this will be done with a gear system; within this category, a worm wheel type transmission (also called worm-wheel) is selected because this type of arrangement can transmit high loads at low speed, which is a desired operating condition. The sizing process will be carried out according to Gieck [40]. The equations used, as well as the meaning of the variables and the results are presented in table 2.1.

Table 2.1. Results of worm and wheel drive sizing

Fórmula	Variables	Datos	Resultados
$R_i = \dfrac{Z}{f}$	Rt: relación de transmisión Z: número de dientes de la rueda. f: número de filetes del sinfin	Z=20 f=1	20:1
$N_r = \dfrac{Ng}{R_i}$	n_r: velocidad de giro de la rueda rpm n_g: velocidad de giro del sinfin o gusano rpm	Ng=6 rpm Rt= 20	0,3 rpm
$p = \sqrt[3]{\dfrac{15 \times P}{n_g \times \sigma_{adm} \times \psi \times f}}$	p: paso del tornillo sinfin p: potencia transmitida, en cv n_g: velocidad de giro del tornillo sinfin, rpm σ_{adm}: esfuerzo admisible del material del tornillo, en kgf/mm^2 ψ: factor que depende del tipo de material f: número de filetes del sinfin	P=0,0004352 cv n_g= 6 σ_{adm}=6,1 ψ= 2,5 (para bronce y aluminio) f= 1	1 mm
$dp = 2 \times k \times m$	dp: diámetro primitivo del tornillo sinfin, en mm k: factor que depende del número de filetes m: módulo de la transmisión	k:=3,4 (para 1 filete) m= 3 (*)	20,4 mm
$de = dp + (2 \times m)$	de: diámetro exterior del tornillo sinfin dp: diámetro primitivo del tornillo sinfin	m= 3 dp= 20,4 mm	26,4 mm
$df = dp - (2,33 \times m)$	df: diámetro de fondo del tornillo sinfin, en mm dp: diámetro primitivo, en mm	dp= 20,4 mm m= 3	13,41 mm

Table 2.1. Continued

$l = 2 \times m \times \left(\left(\sqrt{Z}\right)+1\right)$	*l*: longitud del tornillo sinfín, en mm *Z*: número de dientes de la rueda	*m*= 3 *Z*=20	32,83 mm. Se toma 35 mm
$(^{*})\, m = \dfrac{de}{Z+2}$	*m*: módulo de la transmisión, en mm *de*: diámetro externo de la rueda, en mm *Z*: número de dientes de la rueda	*Z*=20 *de*= 68 mm, dato obtenido por medición de la perilla de accionamiento del tornillo que posee el equipo actual	3,1. se toma el valor 3
$dp_g = m \times Z$	dp_g: diámetro primitivo de la rueda en, mm *Z*: número de dientes de la rueda	*m*= 3 *Z*=20	60 mm
$de = m \times (Z+2)$	*de*: diámetro externo de la rueda, en mm *Z*: número de dientes de la rueda	*m*= 3 *Z*=20	66 mm. Valor calculado considerando módulo igual a 3
$df = m \times (Z - 2{,}33)$	*df*: diámetro de fondo de la rueda, en mm *Z*: número de dientes de la rueda	*m*= 3 *Z*=20	53,01 mm
$\tan \beta = \dfrac{f}{(2 \times k)}$	β: ángulo de avance del tornillo sinfín *f*: cantidad de filetes *k*: factor que depende del número de filetes	*f*= 1 *k*=3,4	8,4°
$\tan \phi = \mu$	ϕ: ángulo de fricción deslizante μ: coeficiente de fricción deslizante	μ= 0,18 (para bronce sobre acero)	10,2°
$\tan \phi' = 1{,}034 \times \mu$	ϕ': ángulo de fricción total μ: coeficiente de fricción deslizante	μ= 0,18 (para bronce sobre acero)	10,54°
$\eta = \dfrac{\tan \beta}{\tan(\beta + \phi')}$	η: eficiencia de la transmisión	β= 8,4° ϕ'= 10,54°	0,43
$F_t = \dfrac{2 \times T_r}{de}$	F_t: fuerza tangencial en la rueda, en N T_r: torque aplicado, en N.m *de*: diámetro externo de la rueda, en mm	T_r= 10,2 *de*= 66 mm	309 N
$F' = F_t \times \tan(\beta + \phi')$	F': fuerza axial	β= 8,4° ϕ'= 10,54	106, 04 N

2.3.1. **Computer-aided design packages**

Computer Aided Design (CAD) packages are applications that provide a range of tools and resources to carry out the design and validation of prototype mechanical systems to be implemented.

Computer-aided design is part of the resources used in mechatronics. An important objective of CAD in mechatronics is to integrate different disciplines throughout the design process [41]; this helps to include mutual interactions between subsystems of different nature and to unify the documentation of all actions and results obtained at each stage of the project.

Generally, in UPTA, the SolidWorks package from the company of the same name and the Inventor package from Autodesk are used.

The former is a fairly powerful parametric design software which, in the latest versions, includes external and internal flow analysis. It is compatible with packages such as Autocad, Pro-Engineer, Inventor. It has compatibility with applications from National Instruments (Control Motion) as well as the Sim Mechanics Link application from The Mathworks company.

The Inventor software is also quite versatile, although not at the level of SolidWorks. Unlike SolidWorks, it is simpler to operate; you can simulate the movement of the designed mechanisms and thus evaluate their functionality. Both SolidWorks and Inventor have CAE analysis modules - another component of mechatronics - and an extensive materials library.

In the development of the present work, the modelling of the components will be done with the Inventor package.

2.4. **Process simulation, data acquisition and processing packages.**

The LabView programme

LabView is a high-level, graphical programming language initially focused on instrumentation applications. LabView programs are called *Virtual Instruments* (VI) because their appearance and operation mimic physical instruments, such as oscilloscopes or multimeters. This language contains a wide variety of tools for data acquisition, display and storage, as well as tools to help specify its execution code [42]. Figure 1.8 shows an example of a virtual instrument made with this package, applied to the handling of fatigue test data performed with the machine developed by GUNT.

The Movicon programme

This package is produced by the Italian company Progea. It is basically oriented towards the design of SCADA systems; one of the fundamental characteristics of such systems is the exchange of data with remote stations. Due to the nature of the system to be redesigned (there is no remote connection to any other apparatus in the UPTA materials laboratory), the LabView package is chosen as the platform for the control and monitoring part of the proposed solution.

User interface

For the development of the interface, the following variables were considered:

- Input variables: desired load, permissible error, rotational speed, current.
- Output variables: number of pulses, applied and de-applied load, generated stress, voltage.

The main function of the interface is to be told the value of the test load and the permissible error. Once this information is loaded, the value of the calculated stress is stored; after the specimen breaks, the value of the number of cycles at which the specimen failed is stored. These data of stress and number of turns are saved in an Excel (*or Spreadsheet*) file, to be later processed with another application.

The redesign of the current system also includes the electrical and electronic part since the ignition, adjustment and shutdown of the motor will be done by means of the created programme; devices (solid state relays, motor rotation direction controllers among others) must be selected to operate the motor before and after the breakage of the test piece and to send this signal to the data acquisition card.

The proposed model retains the current system's probe break detection sensor (shown in figure 2.3). The alternative solution is shown in figure 2.5. The complete machine is not shown here, as the main modifications were made to the load application assembly.

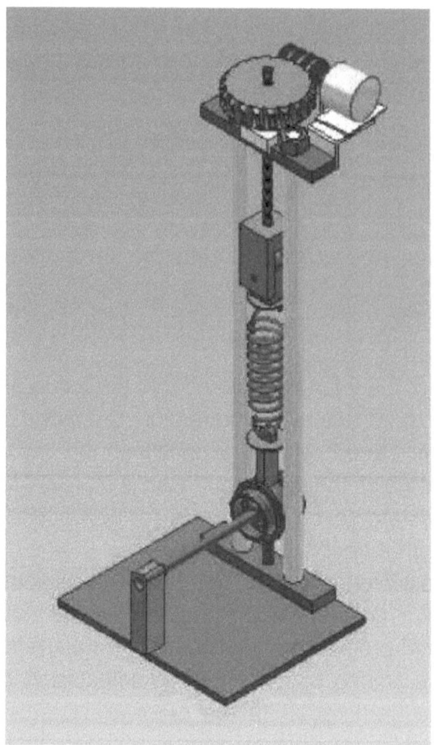

Figure 2.5.Overview of the proposed solution

Figure 2.6 shows a detail of the upper part of the bending load application system, with the drive assembly. The knob is replaced by a worm gear system. The motor support is a generic model (see drawings in annex N° 1). The support element at the front of the test piece is only representative, to give an idea of the assembly. The actual application makes use of the mandrel as shown in figure 2.1.

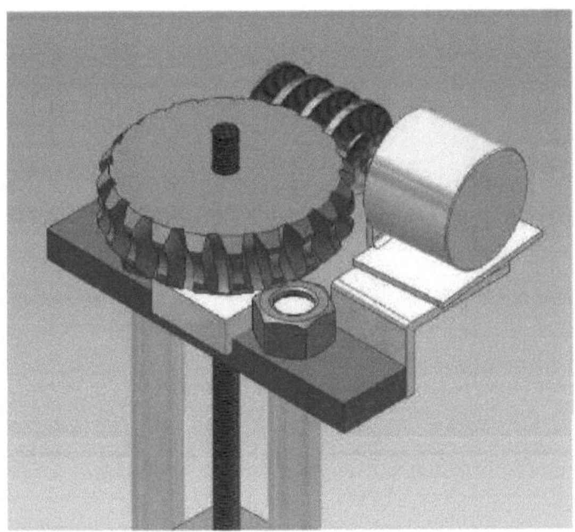

Figure 2.6. Detail of the worm gear system, which replaces the original manual knob system.

The drive assembly is controlled and monitored by the interface generated in LabView. It is shown in figure 2.7. It also has simulator functions. Due to the very long test run time, this feature was incorporated to indicate how the proposed solution operates. It can be seen that it has a built-in history graph, so that in the simulation, the evolution of the applied load can be followed. The operation is as described below:

In the "Desired load" window, the user enters the value of the test load, the bending load. Then, the permissible error value is entered, which is read by the data acquisition system and, through the strain gauge, it senses the force exerted by the load application system. By error comparison, the signal is sent to the drive assembly to stop its action. When the comparison is finished, the user is given the possibility to start or stop the test as shown in figure 2.8.

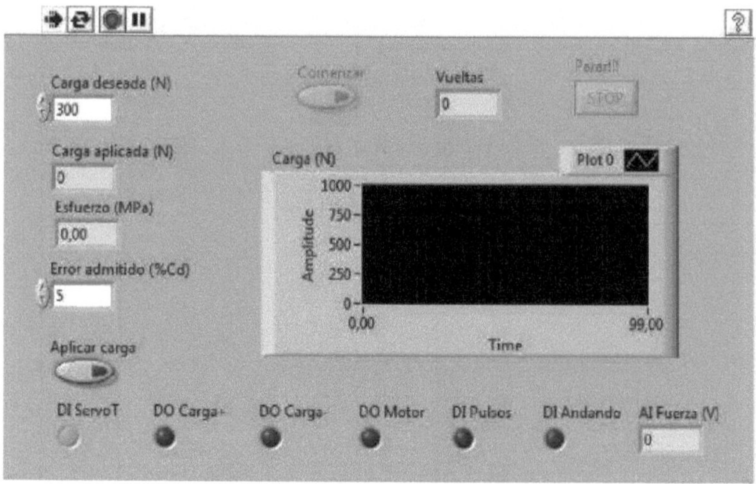

Figure 2.7. Control and monitoring interface of the proposed solution. It also serves as a fatigue test simulator.

The set of lights (LED's) located at the bottom of the interface indicate the monitoring process that is carried out at the same time as the test. It is also observed that, when the error admitted in the application of the load is detected, the stress generated is calculated internally and its value is displayed. This value, due to the type of bending produced, corresponds to the value of the stress amplitude which is considered in the elaboration of the Wohler curve. The evolution of the application of the load can also be seen in the history graph incorporated in the interface.

Annex 2 shows the block diagram with the corresponding coding. It does not include the export process to Excel. Annexes 3 and 4 present the selected data acquisition card and the motor rotation control system (driver).

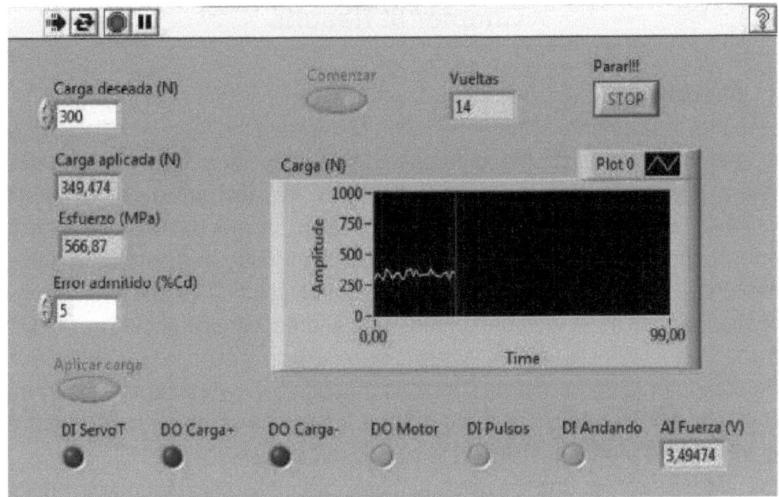

Figure 2.8. Simulator-interface run

2.5. Partial conclusions

The proposed transmission mechanism, of the gear - worm type, is suitable for load application due to its ability to transmit large loads at low speed.

The proposed interface configuration allows monitoring and controlling a real test as well as simulating the process.

To implement the proposed solution it is not necessary to significantly modify the structure of the original model of the fatigue testing machine as can be seen in figures 2.5 and 2.6.

Chapter 3. Results

3.1. Introduction.

In the previous chapters, the importance of knowing the fatigue resistance of the materials of which the different components of a given mechanical system are made was discussed. Some of the technological resources available, as well as their concrete application in mechatronic devices to perform the rotary bending fatigue test, were also presented in these chapters. Additionally, the proposed alternative solution for the problematic situation of the UPTA was presented. This alternative was shown in general terms.

The computer tools for the design of its mechanical components and for the simulation and supervision of the fatigue test that was being carried out at the aforementioned institution were discussed and presented.

This chapter shows the solution in detail with the corresponding design validation as well as an overview of the software verification of the force application system calculation.

Finally, a comparison table between the proposed solution and the current system is presented together with a cost comparison between the proposed solution and a completely new equipment.

3.2. Designed system

As stated in the previous chapter, after reviewing the proposal shown in the above-mentioned reference on how to apply the bending load, it was concluded that it is better to keep the force screw system of the current model as it is quite functional.

Then, the development of the solution was focused on making the drive of the same to be done in a controlled way; based on this premise, it was determined that a gear system driven by a motor controlled through an application made with the LabView program is suitable to apply the load in the desired way and that the gear system that best suits is the worm-wheel type gear or worm-wheel mechanism as it is also known.

The previous estimates led to dimensions, which were determined in the previous chapter, summarised in table 3.1.

Table 3.1. General data of the designed load application system

Wormwheel pitch diameter	20.4 mm
Outside diameter of worm and wheel	26.4 mm
Length worm screw	35 mm
Primitive wheel diameter	60 mm
Outside wheel diameter	66 mm
Number of wheel teeth	
Pressure angle	20°
Transmission module	
Efficiency	43%

With these values, the parts were modelled using the Autodesk Inventor program. The assembled assembly is shown in figure 3.1 (see also assembled assembly drawing in annex N° 1).

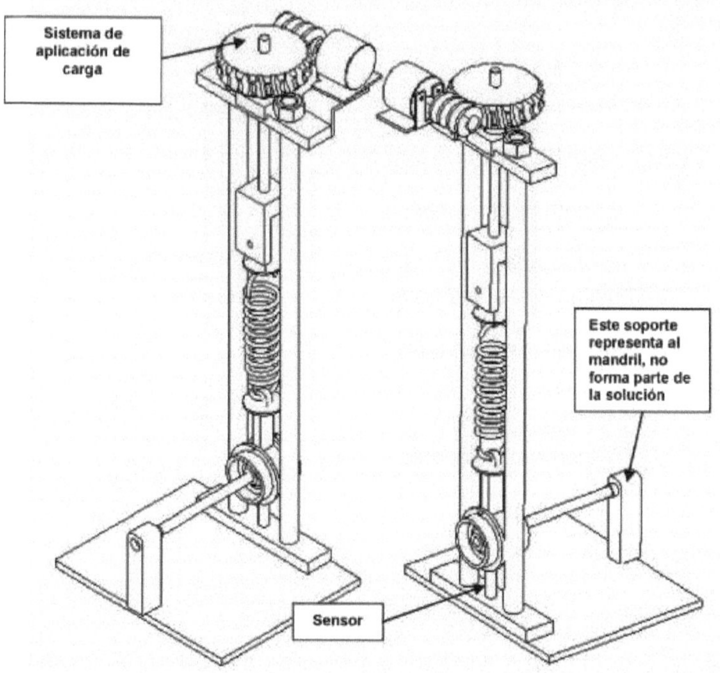

Figure 3.1. Overview of the proposed system showing new elements such as the probe holder position sensor and the worm gear. The strain gauge is not shown.

This figure shows a sensor at the bottom of the specimen holder. Its function is to indicate to the system when the specimen holder has reached its normal position and thus send the signal to turn off the drive motor of the worm-wheel gear mechanism. It consists of a single switch which is held in the normally closed position. When it changes to open, the system interprets the information as a load application and then the motor is kept rotating in one direction by the corresponding controller. When the test specimen is broken - and the test ends - the controller reverses the motor rotation to return the specimen holder to the normal position by returning the sensor to the normally closed position. Figure 3.2 shows the location of the strain gauge that records the value of the applied load.

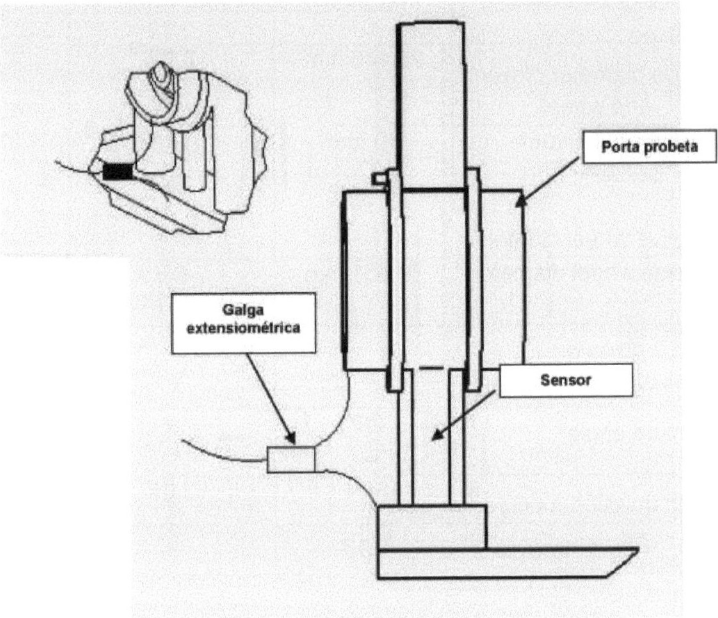

Figure 3.2. Detail of strain gauge location.

The data shown in table 3.1 were checked and corrected using the Design Accelerator tool included in the Inventor program; the characteristics of the designed transmission can be found in annex N° 5. Table 3.2 shows a summary of the specifications of the transmission calculated and dimensioned with the aforementioned tool.

Table 3.2. Summary of specifications of the designed model

		% change from target value
Worm primitive diameter	19,845 mm	2,72
Outside diameter of worm and wheel	26,145 mm	0,97
Length of worm	40 mm	14,3
Primitive wheel diameter	63 mm	5
Outside wheel diameter	70,481 mm	6,8
Number of wheel teeth		-
Pressure angle	20°	-
Transmission module	3,11	3,67
Efficiency	0,43	-

This type of transmission is, in most cases, constructed with two types of materials: bronze for the worm and carbon steel for the wheel. Bronze was chosen for the worm as it is expected to fail instead of the wheel. Although it is a standard transmission, which implies its functionality, the design was validated by finite element analysis assuming that the chosen material withstands the existing working conditions. The designed model also includes a bearing on which the wheel of the worm gear-screw drive assembly is supported in order to further reduce the effect of friction. Such a device is shown in figure 3.3.

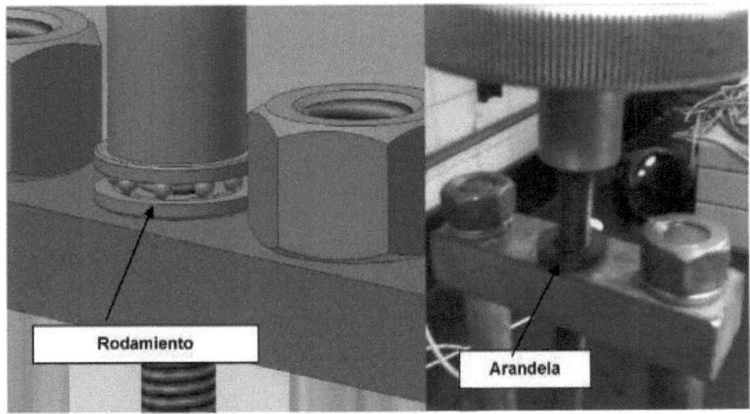

Figure 3.3: Bearing detail in the designed system and configuration of the current system

The Design Accelerator tool generates the torque of the worm-wheel parts, however, as the worm is the most stressed part, the analysis was carried out for this one only. Annex N° 6 shows the results of the stress analysis with the table of values obtained. Figures 3.6 to 3.10 show a summary of the graphical results of this analysis. Figures 3.4 and 3.5 show the applied load and the restriction to movement that were used for the simulation with the CAE tool of the Inventor program.

Figure 3.4. Load applied for
analysis

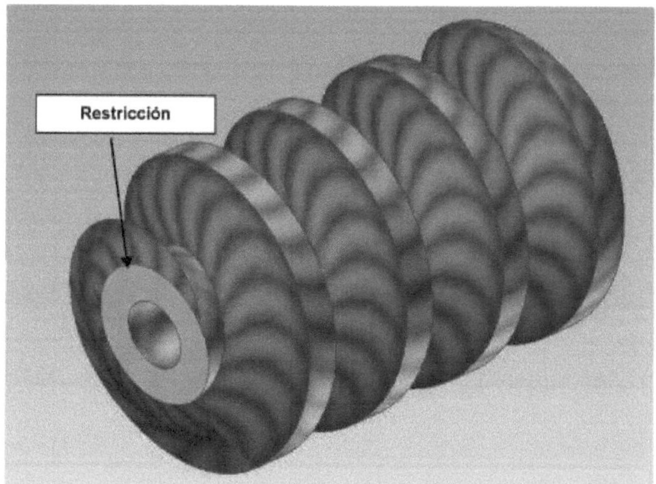

Figure 3.5. Location of the
constraint

The graphical results of the analysis are shown below

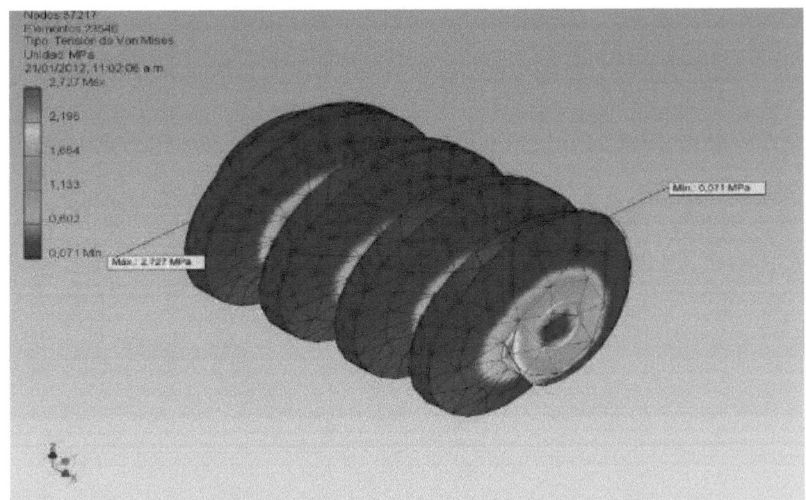

Figure 3.6: Von Mises Stress

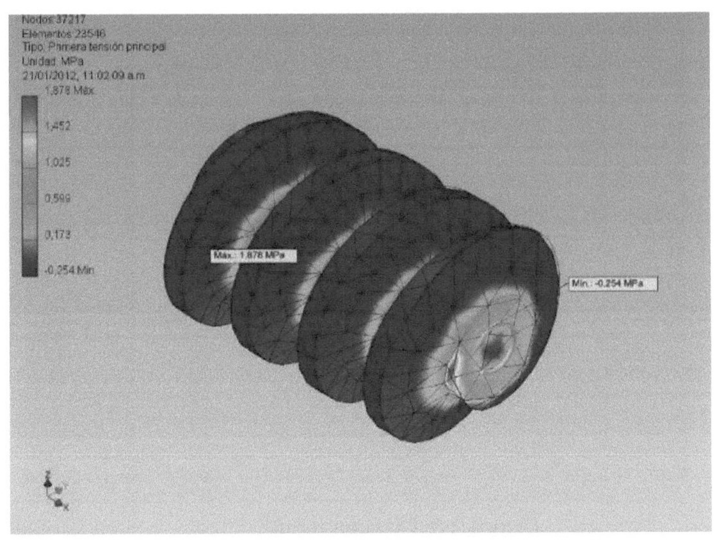

Figure 3.7: First main voltage

It can be seen in this figure that the maximum stress occurs at the start of the worm helix.

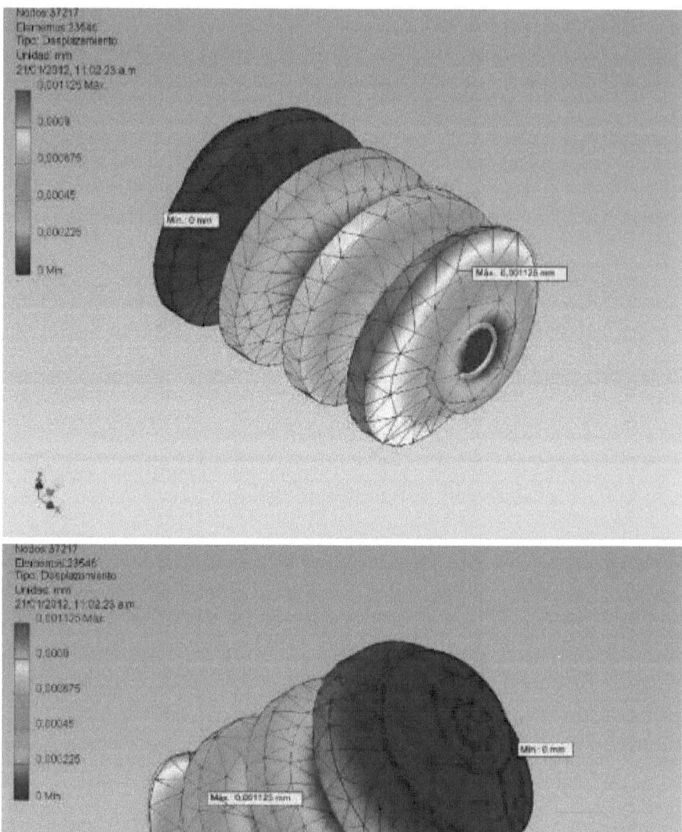

Figure 3.8: Displacement

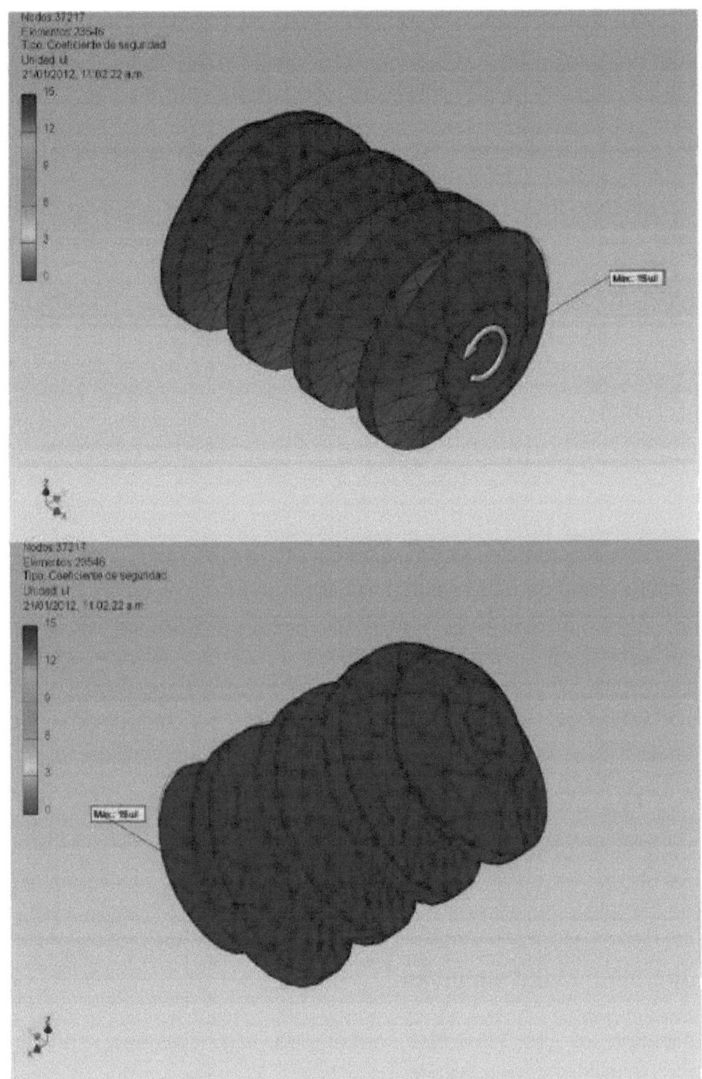

Figure 3.9: Factor of Safety

The mesh validation process was also performed to verify that the default size of the finite element is adequate. The result of this process is shown in table 3.3; two reductions were made to the element in question

Table 3.3. Mesh Validation

Mallado	Maximum stress (MPa)	Percentage error
Default (0.100 mm)	2,668	-
1° Reduction (0,075 mm)	2,563	3,93 %
2nd Reduction (0.070 mm)	2,574	3,5 %

According to these results, the default size of the element is adequate.

As an element for data acquisition and handling, the USB-6008-12 Bit model card from National Instruments is used, the characteristics of which are shown in Appendix No. 3. For controlling the direction of rotation of the motor of the worm gear system, "H-bridge" type devices are suggested. One model is the BD6226 from Rohm & Co. with a voltage of 18 V, output current of 1 A and 2 channels as shown in Appendix No. 4. Both devices are of modern technology and are widely used in the industry, so that, if they are used in the solution proposed in this work, the machine can operate efficiently, which is what is desired of it.

3.2.1. Operation of the simulator

The operation of the simulator, developed with LabView software, is described in general terms below:

On the front panel (see figure 2.7), the desired load and permissible error data are entered.

Looking at the block diagram, the central structure contains the instructions that process these data. It is shown in Figure 3.10.

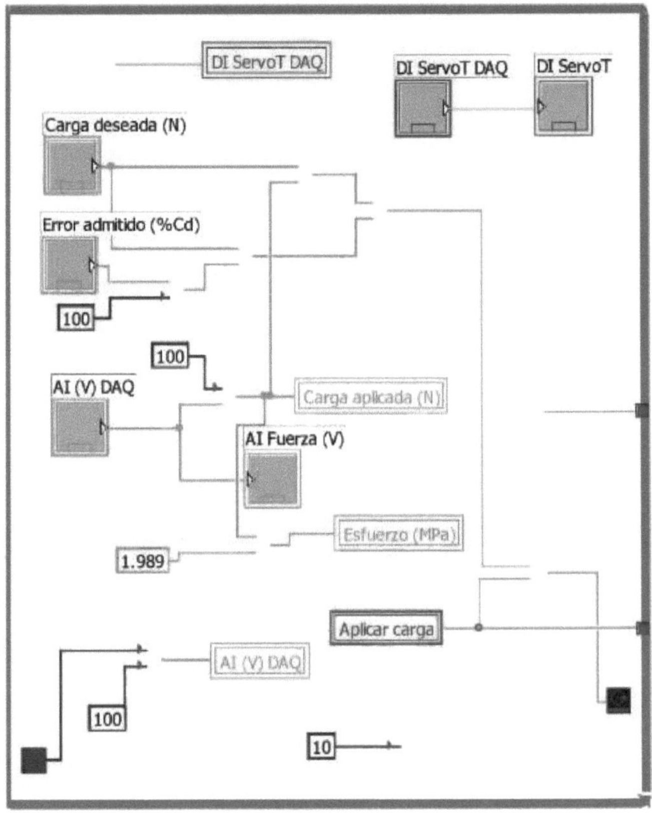

Figure 3.10. Processing module of the desired load value
and the error admitted by the simulator. Source: Author

The "DI Servo" LED is on. At that moment, the probe holder is in its initial position, i.e. it has not moved. This variable is of Boolean type and its control is carried out from the module shown in figure 3.11.

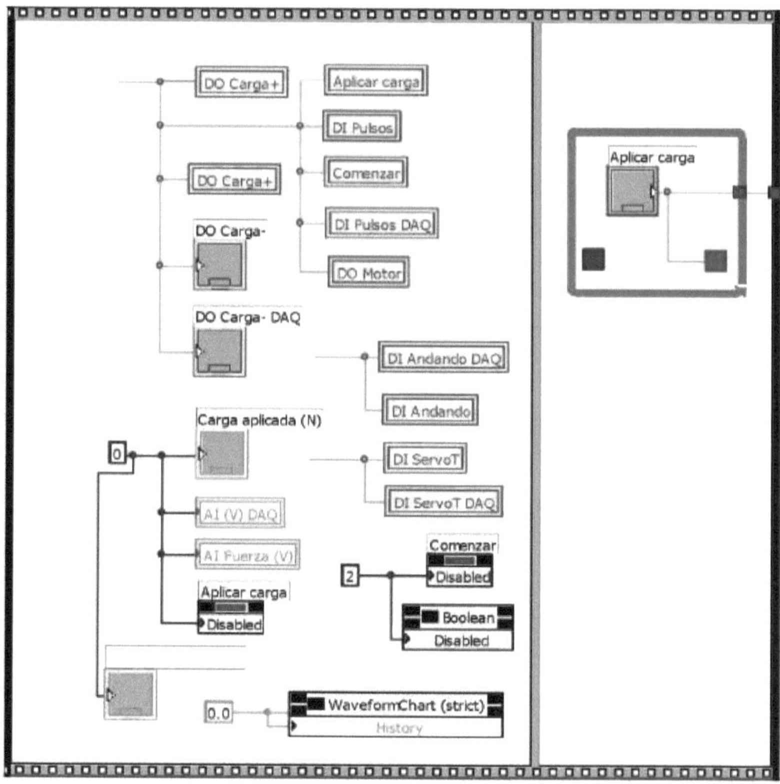

Figure 3.11.
Force system motor on/off control module.
Source: Author

In the module shown in figure 3.10, the actual error calculation process is performed and compared with the desired error. This comparison process is initiated only when the user presses the "Apply Load" button; when this last command is executed, the "DI servo" indicator is turned off and goes to a logic zero (open) state. The normal state of the servo is normally closed (one). Then, almost instantaneously, the "DO load+" LED lights up, indicating that the motor is running.

Once the comparison process is finished, in which the calculated error is less than the admitted error, the led "DO load+" goes off and, as shown in figure 3.10, the corresponding stress is calculated. This stress value is the one that belongs to the stress field and is therefore the test stress. This stress is calculated with the formula given in [11] . The block diagram in figure 3.10 shows that the value of the test load is multiplied by the constant 1.989 to obtain this stress value. The constant is the result of applying the bending stress equation given in the cited reference.

Once the stress has been calculated, the "Start" button is activated which, as shown in figure 3.11, is disabled; parallel to this action, the "Stop" button is activated, which is in the module shown in figure 3.11. This button gives the user the option to stop the simulator run, and also that of the real machine, if required. This stop condition will be explained later.

The start command given by pressing the "Start" button is executed in the module block diagram shown in Figure 3.12 and enlarged in Figure 3.13.

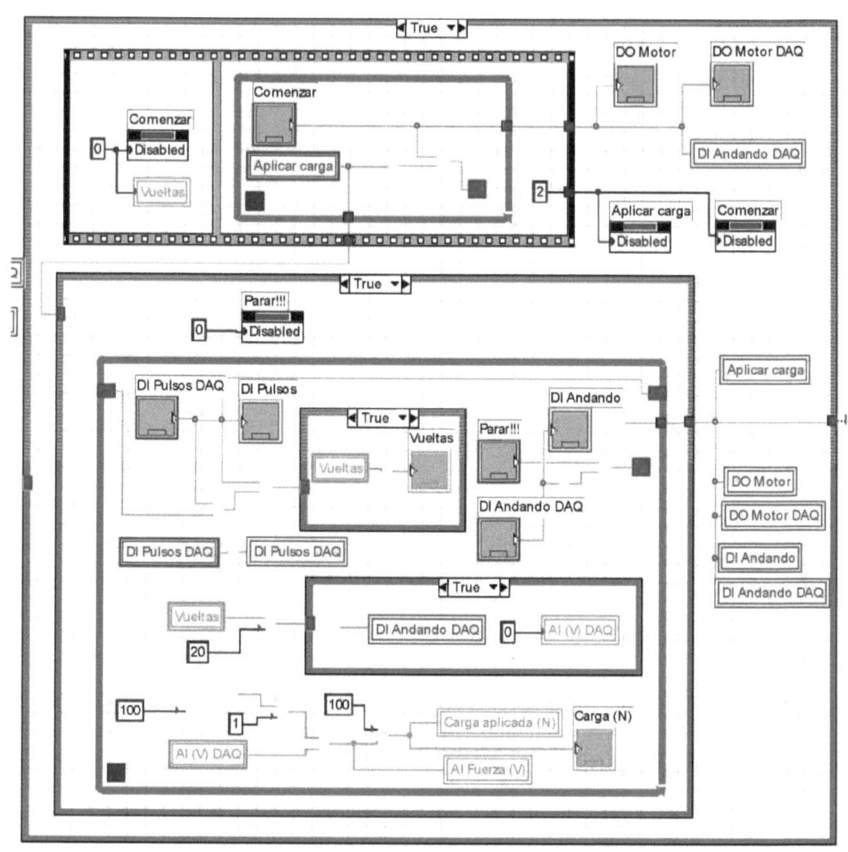

Figure 3.12. Module with the instruction "Start".
Source: Author

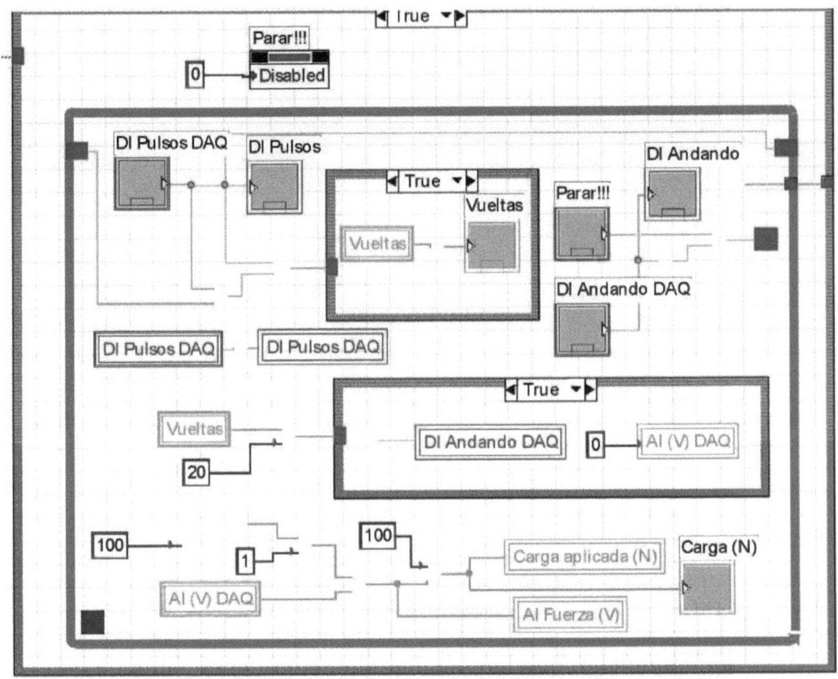

Figure 3.13. Internal loop where the
"Start" instruction is executed.
Note also the CASE for counting the turns of
the test tube until it breaks and the CASE indicating that the
machine' s motor is running. Source: Author

Figure 3.12 shows the sequence structure where, at the beginning, the "Start" button and the output window of the number of laps are disabled. Once it is detected that the error margin is the allowed one, the following sequence is executed. The internal CASE is kept at the value TRUE. In this condition, the variable associated with the "Stop" button is disabled and the FOR loop is executed as long as the "Stop" button is not pressed.

The led "DI Pulses" represents the pulse input registered by the lap counter. The tachometer of the current model is equipped with a pulse counter, and therefore the solution

The same method is proposed for counting the number of revolutions of the specimen until breakage.

It is noted that the internal CASE, which records that the engine is running, is connected to a number of revolutions comparison structure with a constant (20). This is for simulation purposes only. Thus the machine is commanded to stop indicating that the test piece broke. The physical model will take the break signal from the test piece by detecting the voltage variation generated at the strain gauge at that time.

At the bottom of the FOR cycle is the process for generating the curve showing the evolution of the applied load from the start of the machine's engine until the probe breaks and the engine stops. A summation operation is shown between the recorded voltage value, denoted as VI DAQ, and a random number. This result is then multiplied by a constant (100) representing a gain value of the force detector circuit (a value which is assumed, assumed for simulation purposes). The result of the summation is also expressed in an output window to indicate the voltage recorded by the strain gauge. In the real model this whole process is part of the fatigue test monitoring function performed by the application.

The introduction of the random number in the generation of the curve is to indicate that small vibrations occur during engine rotation, which propagate in the load application system, as shown in figure 2.8. These do not represent any danger for the development of the test, unless there could be deterioration of the bearings. All variables marked with the letters DAQ indicate that they are recorded by the data acquisition card specified earlier in this chapter, i.e. they would be the variables of the actual machine.

Now, if the user presses the stop button, the CASE structure in figure 3.13 is set to False and the following occurs in figure 3.14

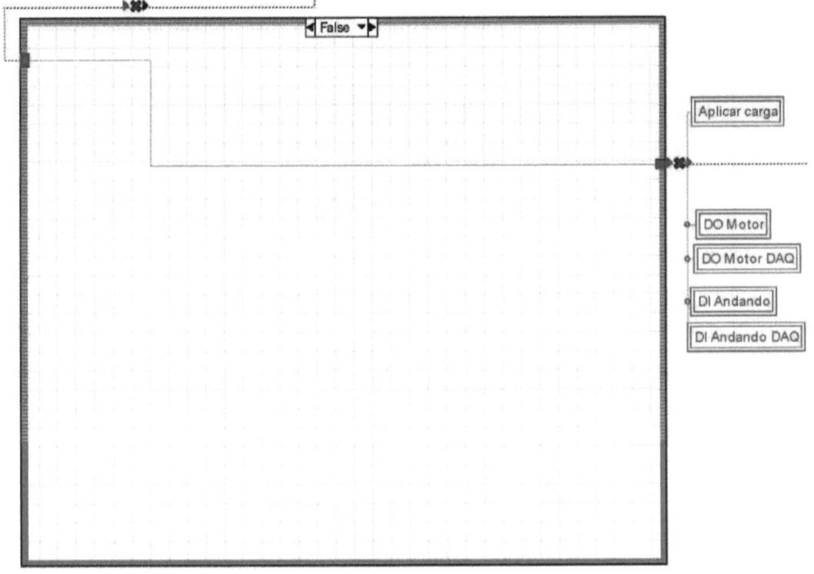

Figure 3.14. When the stop button is pressed, the CASE selector changes
to False and switches back to the
load application situation. Source: Author

In this figure the system is commanded to restart the bending load application
phase of the test.

Finally, with the CASE structures in figure 3.13 set to True, the sequence
shown in figure 3.15 is activated. This controls the motor of the force
application system (see figure 2.6). It can be seen that in the FOR cycle, the
product of the recorded voltage by a gain value (100) is again carried out to
obtain the value of the applied load. This drive is executed in the sequence
shown in figure 3.11, where the value of the Boolean variable is kept at the
value True (T). At this stage, the "Apply load" button is disabled.

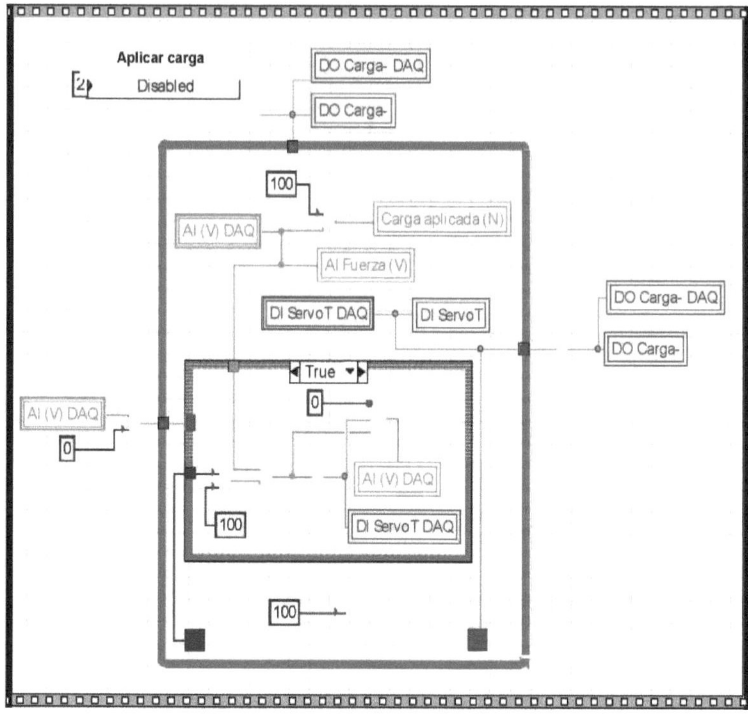

Figure 3.15. Loop sequence for motor control of the
load application system.
Source: Author

This control also allows, after the specimen breaks, to return the specimen holder to its original position causing the "DI Servo" LED to light up again, thus ending the test.

3.2.2. Electronic system

Annex N° 8 shows the circuits corresponding to the current model and the proposed model.

The current circuit has a bridge rectifier that processes the current signal for the pulse detector circuit to process and send to the counter (tachometer). This detector

The pulse sensor has a magnetic sensor attached to the rear of the machine's motor shaft.

The proposed system circuit has a solid state relay, whose function is to automatically start the motor when the error condition in the applied load set in the program is satisfied.

This circuit also has a relay that activates the pulse counter and also a motor rotation direction controller, which works with an H-bridge, as shown in annex N° 4.

3.3. Comparison between the current system and the proposed model

Although the proposed solution retains much of the structure of the current machine, there are differences which are summarised in table 3.4 below.

Table 3.4. Differences between the current system and the designed system

Current system	Designed system
The load is applied manually	The charge is applied by command from the computer.
The value of the applied load is read from the column of the dynamometer. The stress is calculated separately	The values of load, stress and number of cycles to failure are read at the interface of the controller-supervisor system.
Data are collected manually	The test data is saved in a text or spreadsheet file for further processing.

3.4. Approximate cost of the proposal

The following aspects are considered in estimating the cost of the proposal:

- The prices specified are for those components that cannot be produced in UPTA's facilities. The cost of parts produced with the institution's own resources is not specified.
- It also does not include the cost of labour and testing of the load application system already developed. All this is done at the UPTA facilities with the participation of students and teaching staff. Accessories are not included here either (cables, welding supplies, meters, switches, etc.).

- The cost of the solution is approximate, as it is not possible to obtain full component prices.

Table 3.5 shows the prices of the main components.

Table 3.5. Approximate cost of the proposed solution

Component	Cost (US$)
Data acquisition card USB-60008	185
Solid state relay	30
Motor 12 V 6 RPM S330144+mounting base	81
LabView Professional 2012 License	4900
Computer with 2.0 GHz processor	820
Strain gauge	140
Bearing	
H-bridge controller	
Approximate total	**6503,00**

A new machine has a cost of approximately US$ 61700, at the official exchange rate of 4.30 Bsf per dollar (see Annex 7); if the proposal presented in this paper is implemented, there would be a saving compared to the new machine of

$$\frac{61700 - 6503}{61700} \times 100 = 89{,}5\%$$

Savings of 89.5 per cent

3.5. Partial conclusions

The results shown indicate that:

- The dimensions given for the worm screw are suitable for the work to be carried out.

- The material selected for the latter is suitable, as indicated by the stress analyses.

- The addition of the bearing significantly facilitates the operation of the force application assembly by reducing friction.

- The solution is technically feasible from an electronic design point of view due to the commercial availability of the components required for the operation of the proposed design.

- Considerable savings are possible with the implementation of the proposed device.

Conclusions

1. The proposed interface for the UPTA rotary bending fatigue testing machine for the tasks of monitoring, control and data capture with version 7.1 of the LabView package makes it clear that it is not necessary to specify a complex control system. In other words, a higher version of the aforementioned program (containing the motion control libraries) is not required to achieve the desired goal.
2. The interaction or connectivity capability of LabView with other programs allows for rapid processing of the data generated in the test that is the subject of this study.
3. The specified transmission is suitable for moving the bending load application system of the rotating bending fatigue testing machine.
4. The bending load application error detection and adjustment function of the proposed control and monitoring system shown in this paper ensures that no excess bending load is applied to the specimens if the system is built and used for rotary bending fatigue testing.
5. The selected electronic accessories (data acquisition board and motor rotation direction controller), due to their performance, allow the proposed alternative to operate efficiently.
6. The proposal presented in this work, due to its low cost, allows the UPTA to use its resources for other needs that are also a priority, such as the acquisition of metallography supplies, other laboratory equipment, etc.
7. The implementation of the proposed solution presented here will allow practical activities to be carried out to reinforce the students' knowledge of materials, given that materials design and technology are theoretical-practical subjects.

Recommendations

- Use the interaction capability of LabView with other packages to process the test data. Specifically with the Matlab program and the curve fitting tool.
- Explore the possibility of making the assembly of new test tubes and their subsequent disassembly after breakage automatic.

Bibliography

1   **Rodríguez, A.** Evaluation of the fatigue behaviour of 30r steel used in the manufacture of sugar cane mill shafts (Centro de Investigaciones de Soldadura (CIS), Universidad Central de las Villas, Santa Clara, Cuba.

2   **Bangoura, A.** Methodology for the optimisation of high cycle fatigue stress analysis in machine components. *Ingeniería Mecánica, Ingeniería Eléctrica* p. 202 (Instituto Politécnico Nacional, México, D.F., 2007).

3   **Badiola, V.** Variable loads. In Navarra, U.d., ed. *Diseño de Máquinas I* (Department of Mechanical, Energy and Materials Engineering, 2004).

4   **Iberisa**. Introduction to Fatigue or Durability Analysis. p. Fact Sheet No. FT01 (Grupo Iberisa, 2008).

5   **Ripoll, M.L.R.** A statistical fatigue model covering the tension and compression Wohler fields and allowing damage accumulation. *Escuela Tecnica Superior de Ingenieros de Caminos Canales y Puertos,* p. 256 (University of Cantabria, Ciudad Real, Spain, 2008).

6   **Yung-Li-lee, J.P., Richard Hataway, Mark Barkey**. Fatigue.Testing and analysis.Theory and Practice. In Butterworth-Heinemann, E., ed (Elsevier Butterworth- Heinemann, 2005).

7   **Piovan, D.I.M.T.** Theories of dynamic failure. Fatigue failure analysis. In UTN- FRBB, ed2004).

8   **International, E.** Fatigue testing machine. p. 3 (Madrid, 2009).

9   **GUNT.** Fatigue Testing Machine. Hamburg, 2007).

10   **Gere, J.** *Mechanics of Materials.* (Thomson Mexico, 2006).

11   **Spotts, M.F.** *Elements of Machines.* (Pearson Education, 2004).

12   **Ashemi, W.S.J.** *Materials Science and Engineering.* (Mc Graw Hill Interamericana de España, 2004).

13   **col, C.H.y.** August Wohler *Wikipedia The Free Encyclopedia2010.*

14   **Budynas, R.G.** Fatigue failure. In McGraw-Hill, ed. *Shigley's Mechanical Engineering* Design2008).

15   **Federico, C.F.F.** Design and construction of a fatigue fatigue machine by Rotational bending. In SAM, A.A.d.M., ed. *2do Encuentro de Jóvenes Investigadores en Ciencia y Tecnología de Materiales,* p. 6 (Asociación Argentina de Materiales SAM, Posadas - Misiones, Argentina, 2008).

16   **Luis de Vedia, H.S.** Fatigue. *Ensayos Industriales,* p. 4 (University of Buenos Aires, 2002).

17   **Jimenez, G.** Redesign of rotary bending fatigue machine. 2010).

18   **S., F.R.C.** Rediseño y Construcción de la máquina de viga rotatoria sometida a flexión para ensayos de resistencia a la fatiga *Facultad de ingeniería Mecánica,* p. 252 (Escuela Politécnica Nacional, Quito, 2008).

19   **C., M.F.T.** Redesign and construction of a machine for flexural fatigue

strength testing with rotating beam version II. *Faculty of Mechanical Engineering*, p. 304 (Escuela Politécnica Nacional, Quito, 2010).

20   **Instron**. Portable Rotate Beam Fatigue System.

21   **Production, L.d.** Fatigue Protocol. p. 22 (Colombian School of Engineering, 2008).

22   **Commitee, A.I.H.** *ASM International. ASM Handbook.* 1996).

23   **Ramos A., C.E., López Aenlle M., Fernández Canteli A.** Computer application for the analysis of fatigue tests *ANNALS OF FRACTURE MECHANICS,* 2003, **20**, 6.

24   **Bolton, W.** *Mechatronics: Electronic Control Systems in Mechanical and Electrical Engineering.* (AlfaOmega, 2008).

25   **Zhang, H.** *Mechatronic Systems, Applications.* (In-Teh, 2010).

26   **Bishop, R.H. and ISA--The Instrumentation Systems and Automation Society.** The mechatronics handbook. *Electrical engineering handbook series* (CRC Press, Boca Raton, Fla., 2002).

27   **Acar, M.** *Mechatronics in Action.* (Springer, 2010).

28   **Silva, C.W.d.** Mechatronic systems devices, design, control, operation and monitoring. *Mechanical engineering series* (Taylor & Francis, Boca Raton, 2008).

29   **D. Brandolisio, G.P., G. De Corte, J. Symynck, M. Juwet, F. De Bal**. Rotating bending machine for high cycle fatigue testing. p. 72009).

30   **Purdue School of Engineering & Technology, M.E.** Fatigue Test. p. 4 (Indiana University-Purdue University Indianapolis, Indiana, 2007).

31   **Wikipedia.** Servomotor. p. 22011).

32   **Todorobot.** The Servomotor. p. 5.

33   **Ruiz, e.R.** Servos y sistema de Control. p. 10 (Electrónica Empresarial 5IV8, 2009).

34   **Cierzo, C.d.A.** Servos. p. 3.

35   **Robotics, S.** DC motors.

36   **Bishop, R.H.** Mechatronic system control, logic, and data acquisition. (Taylor & Francis, Boca Raton, 2008).

37   **Kreyszig, E.** *Advanced Mathematics for Engineering.* (Limusa - Wiley, 2000).

38   **Kurmyshev, E.** *Fundamentals of Mathematical Methods for Physics and Engineering.* (Limusa, 2003).

39   **Klapper, F.** What is torque and power of an engine and how are they interpreted?

40   **Kurt Gieck, R.G.** *Manual of technical formulas.* (Alfaomega, 2004).

41   **MROZEK, Z.** Computer Aided Design of Mechatronic Systems. *Int. J. Appl. Math. Comput. Sci.* , 2003, **13**(2), 255-267.

42   **Buendía, J.M.M.M.M.J.** *Programming Graphics for Engineers.* Barcelona, 2010)